SpringerBriefs in Spa

Series Editor
Joseph N. Pelton

For further volumes:
http://www.springer.com/series/10058

Ozgur Gurtuna

Fundamentals of Space Business and Economics

Ozgur Gurtuna
Westmount, QC
Canada

ISSN 2191-8171   ISSN 2191-818X (electronic)
ISBN 978-1-4614-6695-6   ISBN 978-1-4614-6696-3 (eBook)
DOI 10.1007/978-1-4614-6696-3
Springer New York Heidelberg Dordrecht London

Library of Congress Control Number: 2013933345

© Ozgur Gurtuna 2013
This work is subject to copyright. All rights are reserved by the Publisher, whether the whole or part of the material is concerned, specifically the rights of translation, reprinting, reuse of illustrations, recitation, broadcasting, reproduction on microfilms or in any other physical way, and transmission or information storage and retrieval, electronic adaptation, computer software, or by similar or dissimilar methodology now known or hereafter developed. Exempted from this legal reservation are brief excerpts in connection with reviews or scholarly analysis or material supplied specifically for the purpose of being entered and executed on a computer system, for exclusive use by the purchaser of the work. Duplication of this publication or parts thereof is permitted only under the provisions of the Copyright Law of the Publisher's location, in its current version, and permission for use must always be obtained from Springer. Permissions for use may be obtained through RightsLink at the Copyright Clearance Center. Violations are liable to prosecution under the respective Copyright Law.
The use of general descriptive names, registered names, trademarks, service marks, etc. in this publication does not imply, even in the absence of a specific statement, that such names are exempt from the relevant protective laws and regulations and therefore free for general use.
While the advice and information in this book are believed to be true and accurate at the date of publication, neither the authors nor the editors nor the publisher can accept any legal responsibility for any errors or omissions that may be made. The publisher makes no warranty, express or implied, with respect to the material contained herein.

Printed on acid-free paper

Springer is part of Springer Science+Business Media (www.springer.com)

For Ela, may her generation solve the destination problem.

This Springer book is published in collaboration with the International Space University. At its central campus in Strasbourg, France, and at various locations around the world, the ISU provides graduate-level training to the future leaders of the global space community. The university offers a 2-month Space Studies Program, a 5-week Southern Hemisphere Program, a 1-year Executive MBA and a 1-year Masters program related to space science, space engineering, systems engineering, space policy and law, business and management, and space and society.

These programs give international graduate students and young space professionals the opportunity to learn while solving complex problems in an intercultural environment. Since its founding in 1987, the International Space University has graduated more than 3,000 students from 100 countries, creating an international network of professionals and leaders. ISU faculty and lecturers from around the world have published hundreds of books and articles on space exploration, applications, science and development.

Foreword

Space business and the ability to create wealth and new markets from off-planet are some of the most important topics for the future of humanity. This book is not only about how innovative people have learned to make profits and establish productive companies in space, but more importantly, it provides a roadmap and guidance for how to construct successful businesses in this important emerging industrial sector. The topic has far-ranging implications for how human enterprise will continue to grow and expand in the future. As we learn how to build safer and more efficient launch systems and the staggering cost of getting mass into space lessens, one can easily imagine new markets arising using space directly or supporting those who do use space. To take full advantage of this opportunity in front of us, it is critical that we understand the lessons of those that went before us in opening the new frontier for commercial enterprise.

At this point in time relatively few space industries have been profitable outside of the government sector. Private industry has had a difficult time exercising its innovative spirit because of the immaturity of the commercial space industry, which has meant there is a high bar to cross to be able to even try potential new product ideas. There are large technology hurdles to overcome and even higher financial mountains to scale. But still, in the face of these obstacles, people have found ways to start profitable businesses. One success that is easy to point to is the space communications industry, which has grown despite these issues and is now estimated to be worth $87 billion. Today many new industries are poised on the cutting edge of success and hold the promise of future profitability; some of these are space resource management, space and suborbital tourism, microgravity materials, and pharmaceuticals manufacturing.

At the beginning of powered flight no one foresaw the airline industry, and at the beginning of mainframe computing no one foresaw the personal computer or the internet and the radical ways these innovations have reshaped our world. Now we stand on a new shore looking out over another unknown ocean of possibilities. Space offers us a limitless region of economic expansion with promises of new solutions to solve old problems and the possibilities of disruptive solutions that will change our world forever. Each year gains are made in lowering launch cost. Each year we learn more about what works and does not work in space through experiments on space platforms such as the International Space Station and from the hundreds of active satellites in orbit. Every success, and every new lesson learned, helps future businesses gain a toehold, lowering the bar to achieve success in space. This book provides the most up to date information on how to take advantage of these new opportunities. This is why this book is so important to read and understand. As is often said, "Do not re-invent the wheel."

Dr. Ozgur Gurtuna understands space business in a way few others can and he has a gift of writing in a style that is easy to understand while tackling the complex topics surrounding space business and economics. He is an entrepreneur having created his own successful company, Turquoise Technology Solutions, Inc., and he knows firsthand how to overcome the challenges of creating and running a business. He taught Business and Management for many years at the International Space University whose central campus is in Strasbourg, France. This experience brought Dr. Gurtuna in touch with the world leaders of commercial space and where he was able to glean the important lessons learned that are in this book. His doctorate is in operations research, where he specializes in risk analysis, optimization technologies, and portfolio selection, giving him a critical eye into how things work in the real world.

It is also fitting that this book is written in partnership with the International Space University, one of the most unique educational organizations in the world. Established in 1987 by three young visionaries – Todd Hawley, Peter Diamandis, and Bob Richards – the university now boasts of over 3,500 alumni from 100 countries. The founders love space and saw in it all the possibilities that the frontier holds. They created the International Space University to be a beacon that draws together all those people from around the world with a great passion for space. Together they explore its endless possibilities and enable humanity's future there. The university hosts an executive training course named the Space Studies Program, which is taught in a new city around the globe each year, and in Strasbourg, ISU teaches Masters Studies with two streams; a Masters of Space Studies and a Masters in Space Business.

Many have tried to create new space businesses in the dawning of this new era, and so far only a very few have succeeded. With the knowledge in these pages you can be part of the new generation of explorers and developers of space.

Strasbourg, France   Gary Martin

About the Author

Dr. Gurtuna is the founder and president of Turquoise Technology Solutions Inc. He is also a faculty member at the International Space University. He obtained his Ph.D. in Operations Research from the joint Ph.D. program in Montreal (this program is administered by four Canadian universities: Concordia, HEC, McGill, and UQAM). His areas of expertise include space business and management, space applications for the energy sector, emerging technology markets and quantitative analysis in decision making (covering areas such as optimization, simulation and mathematical modeling).

Dr. Gurtuna has more than 10 years of strategic consulting experience in the space industry, including experience in strategic planning studies. Previously, he was co-founder and senior consultant of Futuraspace (a consulting company founded by International Space University alumni).

Contents

1	**Introduction to Space Business and Economics**	1
	Physics Enables, Politics Dictates, Economics Sustains	4
	Guide to Contents	5
	Definition of Key Terms	5
	Notes	8
2	**Understanding the Nature of Space Business**	9
	Review of Key Concepts	10
	The Fundamental Forces of Economics: Demand and Supply	10
	Elasticity of Launch Services	11
	Cost Versus Price	11
	The Space "Value Chain"	12
	The Economic Footprint of Space	13
	Satellite Applications: Meet the Three Musketeers	16
	Satellite Telecommunications	16
	Global Navigation Satellite Systems	17
	Remote Sensing	18
	Big Data and Satellite Applications	19
	Notes	20
3	**Seven Distinguishing Features of Space Business**	21
	Cyclical Nature	21
	Linkage to Defense	21
	Government as the Main Customer	22
	The Destination Problem	23
	Limited Competition	24
	Long Investment Horizon	24
	The Curse of the Single Unit of Production	25
	Case Study: Iridium and the Lessons Learned from Terrestrial Competition	25
	Notes	26

xiii

4 Socio-Economic Benefits of Space Activities ... 29
Three Main Types of Benefits ... 30
    Direct Industrial Benefits ... 30
    Spin-offs ... 30
    Societal and Intangible Benefits ... 31
Measuring the Economic Impacts of Space Programs ... 31
A Deeper Look at Spin-offs ... 34
Case Study: On Stardust and Dollars ... 35
Notes ... 36

5 Emerging Space Markets ... 37
Emerging Sectors ... 37
    Space Tourism ... 38
    On-Orbit Satellite Servicing ... 39
    Private Space Exploration ... 40
Notes ... 43

6 Key Issues and Challenges in the Space Business ... 45
High Cost of Access to Space ... 45
Limited Access to Financing ... 46
Inadequate Use of Marketing Tools ... 47
Public Outreach ... 48
Globalization and Consolidation ... 48
The Changing Role of the Private Sector ... 50
Space Debris ... 52
Replacing Generation Apollo ... 53
Notes ... 54

7 Risk Management ... 55
Defining Risk ... 55
Risk in Space ... 56
Types of Risks ... 57
Modeling Risk ... 58
    Risk Matrix ... 58
    Event Tree ... 59
    Hazard Scales ... 60
Managing Risk ... 61
    Testing and Redundancy ... 61
    Diversification: Towards a Portfolio-Based Approach ... 61
    Risk Transfer and Insurance ... 62
Case Study: Nuts and Bolts of Risk Management ... 62
Notes ... 64

8 Cost Management ... 65
Cost Analysis and Management ... 66
A History of Cost Overruns ... 67

	Cost Estimation Methods	67
	Costing by Analogy	68
	Bottom-up Costing	68
	Parametric Costing	69
	Cost of Major Space Programs	69
	Contract Management: The Heritage from the Defense Industry	69
	Cost Management in the Private Sector	71
	Public-Private Partnerships (PPP) as a Cost-Sharing Mechanism	71
	Case Study: Does Anybody Actually Know the Cost of the ISS?	72
	Notes	73
9	**Putting It All Together: Assessing the Feasibility of a Space Venture**	**75**
	Market Overview for OOS	75
	Defining the Business Model	76
	Understanding the Supply Side	77
	Understanding the Demand Side	77
	Is There a Match Between Demand and Supply?	78
	Risk Analysis	78
	Running the 5Ps of Marketing	79
	Pricing	79
	Physical Distribution	80
	Promotion	80
	Product	80
	Philosophy	80
	Notes	80
10	**Conclusions**	**83**
	Top Ten Things to Know About Space Business and Economics	84

List of Acronyms

BRIC	Brazil, Russia, India and China
CBO	Congressional Budget Office
CER	Cost Estimating Relationships
DOD	Department of Defense
ESA	European Space Agency
FAA	Federal Aviation Administration
GEO	Geostationary Earth Orbit
ISS	International Space Station
LEO	Low Earth Orbit
NEO	Near Earth Object
NASA	National Aeronautics and Space Administration

Chapter 1
Introduction to Space Business and Economics

In April 2012, after circling Washington D. C. for a final airborne display, the retired space shuttle *Discovery* landed for the final time. This dramatic last flight marked the end of an era. Having completed 39 successful missions in over 27 years, *Discovery*'s journey ended at the Smithsonian National Air and Space Museum (Fig. 1.1).

While an iconic chapter of U. S. space history was closing, a new chapter was being written. The very next month, the Dragon capsule of SpaceX, a privately held company, successfully completed its resupply mission to the International Space Station, signaling the beginning of private space exploration (Fig. 1.2).

In many ways, our progress in developing space technologies has been remarkably fast. Within 70 years of the first powered flight in 1903, we reached low earth orbit, set foot on the Moon and sent robotic spacecraft to many planetary bodies in our Solar System. These milestones, all achieved within the average lifespan of a human, entirely altered our perception of our species and the Universe around us. The photo of Earth taken by Apollo 8 astronaut William Anders depicted a fragile, blue marble that we call home; a "Pale Blue Dot," in the words of Carl Sagan (Fig. 1.3).

Space investments have brought about very tangible benefits. The most important of these accomplishments include the ability to monitor our changing climate, provide instant communications via satellites, achieve global navigation with pinpoint accuracy, and deploy an ever expanding host of new products and services enabled by our space assets.

The turn of the millennium was not kind to the space sector. The *Columbia* accident hastened the retirement of the space shuttle fleet, and the cancellation of plans for a permanent lunar outpost left a void in future exploration activities. Yet, space business kept on humming between geosynchronous Earth orbit (GEO) and Earth. Satellite-based services such as navigation went mainstream, privately funded spacecraft successfully completed suborbital flights and space continued to be an integral part of our daily lives.

Clearly, the space industry is in a period of transition. The predominance of government as the primary investor and benefactor of space activities is slowly but surely being replaced by a more balanced division of roles and responsibilities

Fig. 1.1 Discovery on its final voyage (Image courtesy of NASA)

Fig. 1.2 The SpaceX Falcon 9 rocket roars into space (Image courtesy of NASA)

Fig. 1.3 "Earth rise" photo taken by Apollo 8 astronaut William Anders (Image courtesy of NASA)

between private and public sector entities. Parallel to this fundamental shift, the number of spacefaring nations is increasing steadily, opening up the "final frontier" of space to more participants than ever before.

Despite all this progress, there are parts of the space industry that are still in their infancy. More than half a century after Yuri Gagarin's historic flight, only a very small group of humans, around 500, have left the gravity well of our planet to reach low Earth orbit (LEO) and beyond. This elite group is composed of professional astronauts (chosen based on their merit) and a few space tourists (chosen based on their wealth). Thus, access to space is still very far from the reach of the masses.

Of course, physically being in space is not a necessary condition to benefit from space, as we will see in the subsequent chapters. The socio-economic benefits from space are largely based on the packages of information we send to, and receive from, space. Nevertheless, as long as access to space is confined to a very low number of humans, the scale of space business will likely be limited. This "destination problem" is one of the main barriers that inhibits the growth of the industry.

Space is a harsh environment not only for human survival but also for business. Despite decades of technological advances and the development of many commercial applications, space business is still very much dependent on government contracts, with few mature segments that can stand on their own. This book is written as a survival guide in this harsh environment as well as an introduction to its many economic and business opportunities.

One of the keys to success in this challenging business environment is to understand how space differs from other sectors of economic activity. Developing viable business strategies is only possible if we take into account the unique nature of space business. By highlighting the specific nature of space business, discussing its many challenges as well as its immense potential, this book aims to be a succinct resource for achieving success in the space business.

This book is written with two main audiences in mind. First and foremost, it is for space professionals who are interested in better understanding the core economics and business concepts applicable to space. Second, it is also for readers who have a business or economics background and a general interest in space.

In a high-tech field such as space, it is only natural that engineering and science are the two dominant disciplines. However, slowly but surely, the nature of space business is changing as a new wave of private investment is flowing into the space industry. The success of these new ventures will be based not only on technical expertise but also on a mastery of business, economics and policy aspects. There are many exciting space ventures targeting different segments of the industry, such as new generations of launch vehicles, sub-orbital space tourism and its related infrastructure, spacecraft for exploring the Moon and other planetary bodies, and even very ambitious plans to mine asteroids. The success of these ventures would not only make space a more prominent part of our daily lives, but it would also cement the role of the private sector as the leading force in space exploration.

Physics Enables, Politics Dictates, Economics Sustains

It is imperative that, no matter their respective disciplines, space professionals need to understand the fundamental role of politics and economics in shaping the past, present and future of the space industry. Just as Newton's laws of motion dictate the orbits of a satellite, the political and economic realities of a certain era determine the types of investments that flow into the space sector. For example, at the height of the space race between the United States and the USSR in 1965, NASA's budget peaked at $28.5 billion (in 2008 dollars).[1] The political will to be the first on the Moon translated into space expenditures of almost 4.5 % of the U. S. federal budget. Based on the 2011 U. S. federal budget, a similar allocation would translate into a NASA budget of more than $150 billion in today's dollars, a far cry from NASA's current budget of $18 billion. And the story is the same around the world. Most industrial nations devote less than 1 % of their resources to space applications, space exploration or space sciences. So what happened after the budget peak of 1965? The political objectives of the lunar landings were achieved, and in the absence of an economic rationale supporting lunar missions, the U. S. administration ended the Apollo program in the early 1970s.

As we will see later in this book, many other space programs suffered the same fate. Building sustainable business ventures that can stand the test of time requires a very solid economic foundation, no matter how strong the initial political motivations may be.

Guide to Contents

In this book, we will tackle many interesting questions, including:

- Why is it so difficult for space ventures to raise private equity or venture capital funding?
- Barring a brief episode of lower costs in the early 1990s, why do launch costs remain so stubbornly high?
- When will the private sector start to operate space missions on their own and on a meaningful scale?
- What are the socio-economic benefits of space exploration?

Economics and business concepts are not sufficient to answer all of these questions (as policy also plays a very critical role), but nevertheless they are essential for understanding the key trends in the industry. More importantly, they can be very powerful as part of our strategic business plans for the future.

In many ways, space is just an expansion of our natural drive to explore, settle and exploit new environments. It's no wonder that exploration and commercial interest have gone hand-in-hand. From the ancient Greek and Phoenician colonies across the Mediterranean to the Age of Sail and the mass settlement of the Americas, economics was a key driver in expanding the footprint of any civilization. Today, the economic sphere of our civilization extends beyond Earth's orbit. It is conceivable that in the future a growing number of planetary bodies in our Solar System will also be centers of economic activity. This feat can only be achieved by a great mastery of technology. However, we also need to develop capabilities based on many other disciplines, such as business, management and economics, in order to make this journey sustainable.

Definition of Key Terms

Some of the key concepts we will use throughout this book are explained briefly below. Readers who are well versed with economics and business may choose to skip this section.

Virtually every product or service in our modern economy is based on one or more "factors of production." These factors include labor, intellectual capital, financial capital, land and natural resources. Every single day, a varying combination of these factors enables the production of millions of goods and services around the world. These outputs include not only consumer staples but also many public goods and services such as education, healthcare and national defense.

Therefore, optimal allocation of these scarce factors is critical to achieving success in business. The disciplines that are dedicated to this purpose are economics and business studies.

Economics is a social science discipline that deals with the allocation of scarce (limited) resources among competing uses, and studies how people make choices to cope with this scarcity.

Markets are the centerpieces of economic activity; they are the "places" where buyers and sellers of goods and services can interact. However, today's markets are no longer attached to a physical location. As the role of the Internet in commerce increases, markets are also being transformed, and virtual marketplaces are formed (i.e., via e-commerce). Traditional markets are complemented and, in some cases, replaced by e-commerce.

Time value of money refers to the very simple observation that "one dollar in my pocket today is more valuable than one dollar next year, even if it will be a guaranteed payment." The combination of various factors such as the opportunity cost of not consuming today, inflation and risk determine the value of money at different time periods.

Discount rate is an adjustment factor that is used to convert future monetary values to today's values. For example, if an individual won a lottery that pays $100 each year, next year's payment would actually be worth less than this year's payment (due to time value of money). If we wanted to find the equivalent value, we would need to decrease next year's payment using a certain discount rate.

Net present value is a commonly used tool in financial analysis that takes into account the time value of money when comparing the costs and benefits of an investment. By using a discount rate, all revenue and expense streams are adjusted to today's values, enabling a more objective assessment.

Base year is a useful concept when we are comparing the value of goods and services in different time periods. Due to inflation and other factors, the value expressed in dollars can change drastically. A base year acts as an "anchor" to adjust the data from all the other years so that meaningful comparisons can be made.

Microeconomics is the branch of economics that studies the decisions made by individuals, households, and firms, and how these parties interact to determine the prices of goods and services and the factors of production. As the name suggests, the level of analysis is at the "micro" level. Thus we study the interaction of individual units. Typical areas of study include consumer behavior and choices, demand and supply interaction and game theory.

Macroeconomics is the study of the entire economic system in terms of the total (aggregate) amount of goods and services produced, total income earned, the level of employment of productive resources, and the general behavior of prices. The analysis is performed at the "macro" level. Here we strive to understand and manage the behavior of the whole system, and not just of the individual parts. Typical areas of study include economic growth, inflation, unemployment, trade balances and fiscal (taxation) policy.

Recurring and non-recurring costs relates to the two key types of costs involved with production. There are many one-time costs during the design, development and manufacturing of a new product. For example, in order to produce a brand new launch vehicle, research and development activities, prototyping, extensive tests and new infrastructure such as launch pads are required. These costs are referred to as non-recurring (or fixed) costs. Once these steps are completed, only the recurring (or variable) portion of the cost remains (such as the raw material and labor needed to manufacture a launch vehicle). The combination of very high non-recurring costs and low production volumes results in very expensive products, a relatively common occurrence in the space industry. Much more profitable and cost-effective results are achieved when non-recurring costs can be spread over millions of production units, such as in automobile manufacturing or consumer electronics.

Business administration/management is the process of leading and directing all or part of an organization, often a business, through the allocation of resources. These resources include all the traditional factors of production (labor, capital, etc.) as well as human capital, intellectual/intangible resources and technology. In management, we deal with the operational, tactical and strategic aspects of leading an organization. Typical areas of study include strategic management, marketing, finance and human resources management.

The gross domestic product (GDP) is the value of all final goods and services produced within a nation in a given year. "Final" is the operative keyword here. In our calculation of GDP we thus consider only the incremental value added by the intermediary steps of producing these goods and services. Otherwise we would significantly overestimate the value of final goods and services. The purchasing power of different currencies can be drastically different (especially for services): it may cost $30 to get a haircut in North America but only $10 in Turkey. Thus, when we compare the living standards of different countries, **purchasing power parity (PPP)** can give a much more accurate picture by adjusting the GDP figures within different countries.

Economies of scale refers to the cost advantages an enterprise can achieve by expanding its operations. These cost advantages can stem from a variety of factors such as buying input materials in bulk, having access to better terms of financing, and dividing marketing and management costs over a greater volume of products and services.

An externality is a cost or benefit incurred by a party, without any compensation, who did not agree to the action causing the cost or benefit. A positive externality occurs when there is a benefit to the party in question, and a negative externality occurs when there is a cost. For example, the economic benefits of spin-offs can be seen as a positive externality, while the threat of space debris can be seen as a negative one.

Demand and supply The quantity of a good or service that is desired by a consumer at a specific price is called demand. Likewise, the quantity of a good or service a supplier is willing to sell at a specific price is called supply. If there is a match between the prices and corresponding quantities, then the market is in "equilibrium," and economic transactions take place. The link between supply and demand is crucial and will be addressed in greater detail in the next chapter.

Systems engineering has been defined by NASA as follows: "Systems engineering is a methodical, disciplined approach for the design, realization, technical management, operations, and retirement of a system. A 'system' is a construct or collection of different elements that together produce results not obtainable by the elements alone. The elements, or parts, can include people, hardware, software, facilities, policies, and documents."[2] Many of the concepts discussed in this book, such as cost and risk analysis, fall under the domain of systems engineering.

In the chapters that follow we will thus learn the scope, the dynamics and the longer term prospects of business in space. We will also explore the economic factors that will influence success or failure in these new space enterprises.

Notes

1. NASA, Aeronautics and Space Report of the President, Fiscal Year 2008 Activities, Washington, D. C.
2. NASA Systems Engineering Handbook, 2007, Washington, D. C.

Chapter 2
Understanding the Nature of Space Business

In November 1984, Dale Gardner and Joseph Allen stepped outside their vehicle to retrieve two broken pieces of equipment. This could have been a very unremarkable task if not for the location of the work: 340 km from the surface of Earth. The mission of Gardner and Allen and their fellow crew members aboard the space shuttle *Discovery* (STS 51-A) was to deploy two telecommunication satellites and to retrieve two other ones that were stranded in LEO. At the time, this commercial mission was seen as a testament to the original vision of the space shuttle – a versatile spacecraft capable of carrying crew and cargo for scientific, military and commercial missions. The "For Sale" sign held by Gardner gave the impression that spaceflight was now a routine, albeit costly activity (Fig. 2.1).

Designed as a multi-purpose vehicle, the space shuttle was indeed open for business in its early days of operation. Between 1981 and 1986 various commercial missions were flown, serving multiple customers. In 1985 alone, *Discovery* flew four separate missions, a remarkable performance underlining the reusable nature of the vehicle. However, the shuttle never became the "space truck" that was originally promised. Just over a year from the successful mission of Gardner and Allen, another mission with a similar designation, STS 51-L, took off from Cape Canaveral. A minute after takeoff, the space shuttle *Challenger* disintegrated midair, taking with it not only the lives of its seven crew members but also the promise of safe and routine human spaceflight.

The *Challenger* accident brought an abrupt end to consideration of the shuttle as a commercial carrier, and many pending commercial launches were canceled. But, what if the *Challenger* accident had not happened? Would the private sector use of the shuttle have been a resounding success? For the reasons we'll explore below, the answer is "Probably not."

Fig. 2.1 Astronaut Dale Gardner holds up a For Sale sign after EVA (Image courtesy of NASA)

Review of Key Concepts

The space industry is markedly different from other sectors of economic activity. Some of these differences stem from the challenging technical requirements demanded by the harsh environment of space. Other differences are related to the origins of the space industry and the strong military rationale that is still an important part of global space activities.

It is not a surprise that the early days of space exploration were dominated by policy imperatives and the intense competition of the Cold War. However, as the space "firsts" were claimed one by one, starting with the world's first artificial satellite, Sputnik, followed by Yuri Gagarin's historical flight and by Neil Armstrong's footsteps on the Moon, the political momentum had slowed down. Today, national pride still plays a significant role, but justifying the economic benefits of space investments in practical terms is becoming more and more important. Thus, mastering the key concepts of space economics helps us to understand the "economic rationale" of space exploration.

The Fundamental Forces of Economics: Demand and Supply

Our review of core concepts will start with demand and supply. It is hardly possible to exaggerate the importance of these two forces that shape the nature of any industry. The formal definition of demand in economics is the quantity of a good or service an individual or group desires at a given price. Supply, on the other hand, is the quantity of a good or service that an individual or group is willing to provide at a given price. Although most people are introduced to the concepts of demand and supply as part of their formal education, very few make use of these concepts to understand the actual behavior of their economic surroundings.

Very simply put, price acts as a lever that determines the actions of buyers and sellers. If the price falls, the quantity demanded rises and the quantity supplied falls. If the price increases, the quantity demanded falls and the quantity supplied rises. This continuous adjustment in the market shapes the quantity and variety of goods and services available. If supply exceeds demand, producers' inventories will increase, forcing them to reduce the price they are willing to accept until supply and demand are equal again.

Sound economic policy and successful business decisions are all based on a careful analysis of the trends that affect the supply and demand for space-related goods and services.

Now let's put these concepts to work and apply them to the space industry.

Elasticity of Launch Services

The sensitivity of consumers to a change in the price of a product or service is measured by the price elasticity of demand. If the demand is inelastic, changes in the price don't have a big impact on the quantity demanded (e.g., cigarette consumption, drinking water). If the demand is elastic, changes in price have a significant impact on the quantity demanded (e.g., hospitality services, luxury items, electronics).

One of the fundamental cost drivers in the space industry is the cost of launching payloads to LEO and beyond. In order to compare the cost of different launch vehicles, one convenient metric is cost per pound or cost per kilogram. A study conducted in 2005 concluded that the oversupply in the launch vehicle market in the last two decades resulted in significant price drops, in some cases as much as 50 %.[1] This price signal would normally cause the demand to increase; however the demand was stable. One explanation for this lack of "demand response" is the long lead times associated with developing new payloads and building spacecraft. Just because there is a cheaper launch available doesn't mean that there will be spacecraft ready for launch. The other, perhaps more fundamental reason, is the limited amount of demand for launch services even in the best of times.

Cost Versus Price

At first glance, the concepts of cost and price are deceptively simple. The former is the total amount of expenses incurred for producing a good or service while the

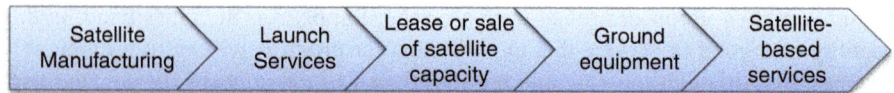

Fig. 2.2 The value chain of satellite applications

latter is the amount a customer is prepared to pay to acquire this very same good or service. Most importantly, from the seller's point of view, the difference between the price she charged and the costs she incurred is the profit – the lifeblood of a market-based economy. Understanding the distinction between the cost and price of a product or service isn't always easy in the space sector, since there is a strong heritage of cost-plus contracts. These types of contracts entitle the contractor to a total reimbursement of all the project costs plus a certain amount of profit, generally based on a percentage of the cost base. Thus, historically, there has been little incentive to control the project costs. It is important to note that the most "commercial" of the space industries, namely the commercial communications satellite industry, was among the first to migrate away from cost plus contracts. By demanding a fixed price bidding process whereby competing satellite manufacturers assumed the risk of cost overruns, satellite operators successfully moved away from a cost-plus business model.

For a private enterprise in a competitive market, the cost of producing a good or service has to be below its market price, so that the firm can make a profit and stay in business. For most government-run projects, the same constraint doesn't apply, so the governments can keep subsidizing very expensive projects because of other priorities, such as national security.

The Space "Value Chain"

Just like any other sector of economic activity, products and services in the space industry reach their final customers after successive rounds of inputs from contributors. In economics, the term "value-added" refers to the additional value created at each phase of production, as raw materials are transformed into finished goods and services by applying factors of production (e.g., labor and capital) (Fig. 2.2).

This value chain may be entirely transparent to a viewer enjoying a live sports broadcast on satellite TV, but it has to operate flawlessly for the customer to enjoy reliable and affordable service. The broadcast signals, coming all the way from GEO, are provided by a TV content provider who has leased satellite capacity from a satellite operator. The operator ensures that the satellite functions optimally by maintaining its orbit and various subsystems. In turn, satellite operators are dependent on many suppliers as well. These include ground equipment manufacturers, satellite manufacturers and launch service providers. Their collective effort is required to design, test, manufacture and deliver satellites in the proper orbit.

Meanwhile, all of these activities are supported by hundreds, if not thousands, of other firms who provide all the necessary hardware and software elements as subsystem suppliers. Unbeknownst to TV viewers who flip on a television set, each satellite-based broadcast from space reaches their living room thanks to this value chain.

The value chain concept applies to all satellite applications, from remote sensing to satellite navigation. Since the value chain captures every essential contribution of the industry participants, it is a very useful gauge to assess the economic activity in the space industry. By summing up all the value-added products and services in each industry segment, we can have a fairly accurate idea about the size of each segment.

The Economic Footprint of Space

Just as in any other branch of science, measurement plays a key role in economics and business, as we cannot truly understand what we cannot measure. For many years, it has been particularly difficult to find reliable and detailed statistics on the space industry. This situation improved drastically in the last few years. Today, some of the leading sources of information that compile industry statistics on the space industry include the following: (i) the Global Forum on Space Economics of the OECD, (ii) various reports published by the Satellite Industry Association in the United States (as compiled by the Futron Corporation), (iii) "Industry Facts & Figures" published by Eurospace and (iv) "The Space Report" published by the Space Foundation. It is interesting to note that global industry statistics are not always in agreement, and cross-checking key indicators is always a good idea.

The inconsistencies may be a result of various factors. One issue is the time period for which the statistics are compiled (i.e., calendar year versus fiscal year). Another issue is the difficulty in combining the figures of largely "wholesale" suppliers such as Intelsat and Eutelsat with the sales figures of retail suppliers. The line between a space-based service and a terrestrial one is not always clear. In some satellite services, such as in broadcasting and telecommunications, for instance, it is not exactly clear where the satellite service has transitioned to a terrestrial telecommunications service or an Internet transaction. Finally, there can be double counting or improper accounting of revenues in a supply chain.

In this section – despite these difficulties – we'll attempt to illustrate the economic footprint of the space industry. As we'll see, when we only consider economic metrics, space is neither a giant industry nor a fringe one. However, some of the most important benefits of space activities are also the ones hardest to quantify. Without key satellite services, thousands of lives might be lost due to storms or hurricanes, airline operations would not be as reliable, financial institutions would not be able to function as efficiently and the reach of the Internet would be severely curtailed – particularly in a number of developing countries. The importance of satellite services ripple across the global economy.

Some of the main factors that determine the size of the space industry are government space budgets. These budgets can give us a good idea about the inputs to

Fig. 2.3 Distribution of satellite services revenues (in billion dollars) between the three main types of satellite applications in 2009 (Source: OECD 2011)

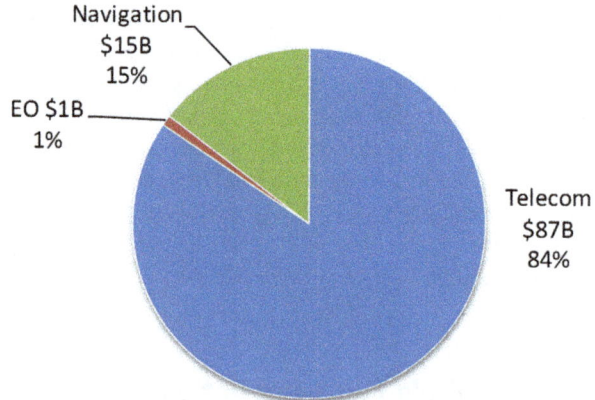

the industry. The OECD regularly publishes reports on government space budgets around the world. The 35 spacefaring nations covered by the OECD invested a total of $64.4 billion in 2009, and an estimated $ 65.3 billion in 2010, including both civilian and military spending.[2] The Space Foundation also publishes statistics for worldwide government space budgets with an estimate of $87.12 billion for 2010. The discrepancy is mostly due to the way U. S. space budgets are calculated, the restricted release of information regarding military satellite activities and lack of standards when it comes to the definition of "space activities".[3]

It is not an exaggeration to say that the United States dwarfs other nations when it comes to space spending. For each dollar spent by the rest of the world on space activities, the United States spends more than $2 (including both civilian and military spending). However, budgetary amounts can be deceiving when making international comparisons regarding the capability of different spacefaring nations. Labor is by far the biggest expense category in space programs, and labor costs are much lower in BRIC countries and other emerging space nations. Therefore, one needs to always check if purchasing power parity (PPP) is used when international budgets are compared.

Another key indicator of space industry statistics concerns the revenues of the commercial sector. Although commercial revenue estimates vary significantly, comparing various sources gives us a range of $170–190 billion for 2010[4,5] This figure includes the combined annual revenues of satellite applications (satellite telecommunications, Earth observation and satellite navigation) and the rest of the value chain (i.e., satellite manufacturing, launch services, ground equipment and support services including launch insurance) (Fig. 2.4).

Tallying up the revenue estimates of the commercial sector can be tricky. It is very easy to double count various revenues. Thus, it is critical to map out the value chain and account for the incremental revenues from one stage to the other properly. Otherwise we can easily overestimate the total volume of the industry. For example, a satellite manufacturer generally subcontracts the many subsystems that go into a satellite. If we tally up the cost of all subsystems to the prime

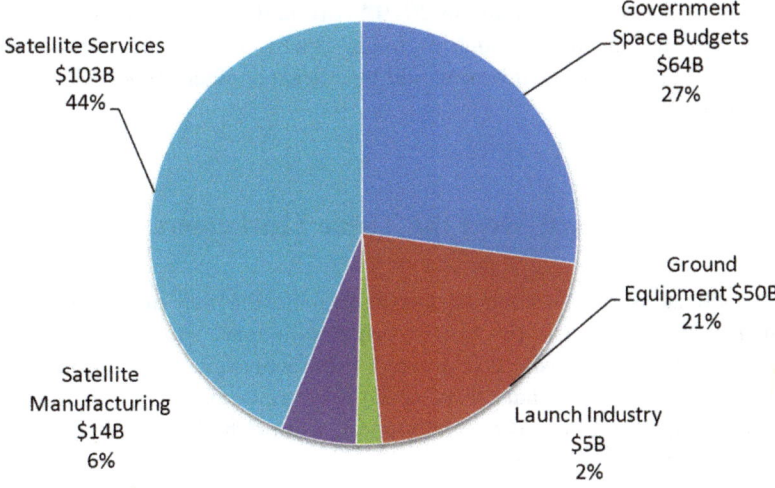

Fig. 2.4 The economic footprint of the space industry in 2009 (Source: OECD and SIA)

contractor and then add the total value of the satellite on top of that, we will be double counting the total value.

The breakdown of the commercial space sector revenues also provides some interesting facts. Satellite applications are by far the leading segment in the space industry, and about 84 % of the aggregate revenues come from a single application, satellite telecommunications. In recent years, the satellite navigation segment has grown at a rapid pace, as location-based services started gaining more importance. In terms of growth rates, direct-to-home broadcast and satellite navigation segments are the two leading segments. Earth observation (EO), a relatively mature segment of the space industry, accounts for about 1 % of the total commercial sales (Fig. 2.3). (This figure does not include the EO products and services purchased by governments).

In order to get a global estimate of the direct economic value of the space sector for any given year, we can add up the public space budgets and private sector revenues. Of course care is needed to avoid double counting. A good portion of the public space budgets flow to the private sector in the form of contracts. For 2010, this global figure encompassing all space activities was in the range of $235–277 billion, depending on which sources we use. This figure excludes indirect benefits (such as spin-offs) as well as the book value of the assets in the sector (e.g., launch pads). So, to sum up, the global economic footprint of the entire space industry is in the range of a quarter trillion dollars.

The workforce of space professionals can also be seen as a key indicator of the global space industry. OECD reports that about 170,000 employees work in space manufacturing in the United States, about 31,000 in Europe and 50,000 in China.[6] Just like for space budgets, different sources don't always agree on these numbers; for example the Space Foundation reports that the U. S. space workforce was composed of 250,000 individuals in 2010.[4] Although it is harder to get statistics for other spacefaring nations, such as India, Brazil, Turkey and other emerging economies, a conservative estimate would be at least another 50,000–70,000 space professionals.

Satellite Applications: Meet the Three Musketeers

Satellite applications have become so ubiquitous today that they may be the victim of their own success. Various types of infrastructure in space support many activities in our daily lives that we easily take granted. From withdrawing cash from ATMs to regulating traffic lights, signals from space drive millions of transactions in our economy. Our quality of life is critically dependent on the flawless operations of a thousand satellites circling the globe.

Satellite Telecommunications

In 1945, a young British engineer published an article in *Wireless World* magazine proposing to cover the entire globe using three telecommunications satellites. This elegant concept, based on these "stationary" orbital points in the sky with respect to the rotating Earth below (GEO orbit) spurred a brand new industry: satellite telecommunications. Sir Arthur C. Clarke's vision, not taken very seriously when it was introduced, became a reality within 20 years with the launch of the experimental commercial satellite, Telstar, in 1962 and the launch of Intelsat I Early Bird (the first commercial geostationary communication satellite) in 1965. Since then, especially GEO – sometimes called "Clarke orbit" in honor of Sir Clarke – has become prime real estate in space, with hundreds of commercial satellites now strategically placed to cover the globe with their signals.

Since the early 1990s, the private sector has been evolving into a more prominent role. At the height of the dot.com boom, satellite telecommunications saw an unprecedented amount of interest for ambitious projects involving hundreds of new satellites, especially in LEO. Mobile satellite telecommunication systems such as Iridium, ICO, Globalstar, and Orbcomm, plus proposed broadband systems such as Astrolink and Teledesic, attracted billions of dollars of private investment. These projects envisioned a dense constellation of satellites providing mobile telecommunications services on a global basis. Unfortunately, just like the sudden rise of these projects, their collapse was also spectacular. Many of them never moved beyond blueprints, and only a few made it to orbit. Some of these new mobile satellite systems have managed to survive, but as a shadow of their original vision. These systems went through bankruptcy proceedings and were typically bought on a distress sale basis for a small fraction of their original valuations.

These past episodes have made it very difficult for space ventures to raise funding, and the drama in the capital markets seems far from over. More recent examples are two companies based in the United States, LightSquared and Terrestar, which planned to combine satellites with broadband mobile terrestrial services as a "hybrid network" principally for the U. S. market. After spending billions of dollars building and launching some of the most massive and sophisticated communications satellites ever conceived, both of these companies have declared bankruptcy. Only the Inmarsat and Thuraya ventures, employing geosynchronous satellite technology, has managed to maintain consistently profitable mobile satellite operations for a sustained period of time.

In stark contrast to the mobile segment, fixed satellite services evolved into a stable and profitable business. This business model is akin to commercial real estate. Satellite operators lease capacity to content providers and telecommunications companies on a long-term basis. The leased capacity is used to provide a host of services, including TV and radio broadcasting and long-distance telephony.

Global Navigation Satellite Systems

Global Navigation Satellite Systems (GNSS) is the generic term that refers to satellite navigation constellations in Earth's orbit. These constellations provide accurate positioning, navigation and timing information to users around the globe.

Currently, there are only two systems that offer global coverage: the U. S. Global Positioning System (GPS) and the Russian GLONASS system. Although other countries have operational navigation satellites, such as the European Galileo, the Indian Regional Navigational Satellite System, the Japanese Quazi-Zenith Satellite System and the Chinese BeiDou/Compass systems, none of these systems provides global coverage at this time.

Originally designed as a military navigation system, satellite navigation has branched out to business-to-business applications (such as surveying) and business-to-consumer applications (such as car navigation). In recent years, the widespread adoption of smart phones with built-in satellite navigation capability has dramatically increased the number of people who regularly use GPS signals as part of their daily lives. The timing function is also vital for a number of scientific and governmental applications. This precision timing capability, enabled by the on-board atomic clocks, has also been used for various applications such as security verification.

Since the GPS signals are provided for free, developing a business case around the space segment is fiendishly difficult. Thus, business opportunities are to be found elsewhere, closer to the end user (also known as the downstream part of the market). Not surprisingly, the lion's share of the GNSS revenues comes from the sales of receivers and associated services to end users. The development of competing GNSS systems is thus largely due to national security reasons rather than an economic rationale. In fact, one can argue that there will be an oversupply of satellite navigation signals once the European, Japanese, Chinese and Indian systems become operational.

A key milestone in the evolution of the satellite navigation was in May 2000, when the U. S. federal government disabled the Selective Availability feature, instantly boosting the precision of the GPS signals for civilian users. In 2007, the U. S. Department of Defense, which procures and operates the GPS satellites, permanently disabled the intentional degradation of the satellite signals.[7] This policy decision significantly reduced the uncertainty surrounding high precision GPS signals for civilian and commercial applications. It has big implications for critical operations such as air traffic control, and it is likely to increase the adoption of GPS-enabled operations in many economic sectors.

Europe is currently deploying its own satellite navigation system, Galileo. When it becomes fully operational towards the end of this decade, Galileo will provide high-precision satellite navigation capability on a global basis.[8] Galileo is designed to be interoperable with GPS and GLONASS and, unlike the GPS, it will be entirely under civilian control. When Galileo joins GPS and GLONASS as an operational satellite navigation system, users will have access to more than 75 satellites.

Certainly such a level of coverage will create a well-developed and reliable upstream segment for the market with a high degree of redundancy. This would virtually guarantee the integration of satellite navigation and timing-based products and services in our daily lives, including in sensitive application areas such as air traffic control and driverless cars.

Remote Sensing

Remote sensing, also known as Earth observation, is one of the most important space-based capabilities at our disposal. Today, thousands of spaceborne instruments are orbiting around Earth, taking the pulse of our environment with precise measurements across the electromagnetic spectrum.

Holding the higher ground has always been a strategic objective throughout history. Successive technological developments such as observation towers, balloons and aircraft were employed to keep an eye on our surroundings. In this sense, remote sensing is nothing new. However, access to space, coupled with advances in electro-optics has opened up a wide variety of orbits around the Earth and carried us to the ultimate higher ground.

The bits of information flying back and forth from these orbiting satellites to Earth provide crucial scientific, military and commercial capabilities. Some of the primary applications of remote sensing include agriculture (e.g., crop classification), forestry (e.g., monitoring deforestation), geology (e.g., mineral exploration), hydrology (e.g., flood mapping and monitoring), meteorology (e.g., numerical weather prediction) and security (e.g., missile launch detection).

On the commercial front, many companies are actively working on developing innovative products and services using remote sensing data. Once the exclusive domain of top secret government operations, high-resolution optical systems are now widely available for commercial applications. Access to high-resolution remote sensing data at a reasonable cost has rapidly increased the scope and variety of these products and services. For example, by using satellite imagery, Remote Sensing Metrics, a U. S. company, is able to create economic activity indicators that are then used as part of economic forecasts. These indicators cover many different economic sectors, such as the number of cars in the parking lot of a shopping mall, the number of shipping containers stockpiled in a port, or the height of the storage tanks in refineries. By observing the changes in these indicators over time, remote sensing analysts can literally see economic cycles in action.[9]

Big Data and Satellite Applications

Another recent development that vastly expanded the use of satellite applications in business is "big data." Although there is no formal definition of big data, it largely refers to our ability to collect, store and analyze vast volumes of data and identify patterns and underlying trends in our economic, natural and political systems.

Vast reductions in the price of digital storage devices, the emergence of cloud computing and virtualization of computer systems have enabled the emergence of big data. A recent report published by McKinsey notes that the amount of data in the world has been exploding, and analyzing large volumes of data will become a central theme of the twenty-first century economic enterprise, unleashing new waves of productivity growth and innovation.[10]

Together, the three main satellite applications collect terabytes of data every day from a wide variety of sensors. Analyzing this "satellite big data" and combining it with other sources of information can create many new business opportunities.

One remarkable example is the Climate Corporation, a company founded by two former Google employees. This company acquired 60 years of crop yield data from the Department of Agriculture, including terabytes of information on soil types for the entire United States. Then, they merged this information with weather forecasts and other climate data from the National Weather Service to calculate the weather-related risks for corn, soybeans and winter wheat. This combined capability enabled the company to track these risks on a continuous basis and develop weather insurance products for farmers.

Another interesting example comes from the solar energy sector. Availability of solar irradiance at a project site is one of the primary factors which affect the profitability of a photovoltaic project. Solar irradiance, in turn, is affected by various atmospheric effects, such as cloud cover, water vapour and aerosols. Therefore, determining the characteristics of these effects at a project site is essential for conducting a profitability analysis and tracking the performance of a photovoltaic investment over time.[11] The imagery from meteorological satellites is very useful for this purpose.

Meteorological satellites, perched in GEO, such as the GOES series of NOAA or the METEOSAT series of EUMETSAT, are capable of imaging the entire disk of

Earth at frequent intervals. Typically, these satellites carry imagers operating at the visible and thermal infrared bands, ideal for detecting clouds.[12] In order to conduct a 10-year historical analysis of solar irradiance, more than 100,000 images need to be analyzed. Today, thanks to advances in data storage and processing power, this type of analysis can be conducted within hours, turning raw satellite imagery into insights for business decisions.

Notes

1. Hertzfeld, H. R., Williamson, R. A., Peter, N., "Launch Vehicles: An Economic Perspective," Space Policy Institute, The George Washington University, 2005 available at http://www.gwu.edu/~spi/assets/docs/NASA%20L.Vehicle%20Study%20V-5.pdf.
2. OECD (2011), The Space Economy at a Glance 2011, OECD Publishing. doi: 10.1787/9789264111790-en.
3. OECD indicates that the U. S. space budget was $43.6 billion for 2010, while the Space Foundation pegs the number at $64.63 billion. It is important to highlight the limitation of some of the figures at our disposal. The Federation of American Scientists, which keeps track of military and civilian space activity, states that tracking the U. S. Department of Defense space budget is very difficult since space projects are generally not identified as separate items. Furthermore, the DOD sometimes releases only partial information or retroactively changes budgetary figures of previous years. NASA's "President Report" gives us a glimpse of the breakdown of the U. S. government's space expenditures across different departments, including the DOD.
4. The Space Foundation (2011), The Space Report, Colorado Springs, CO.
5. Satellite Industry Association (2011), 2011 State of the Satellite Industry, Washington, D.C.
6. OECD, " The Space Economy at a Glance 2011", Paris, France.
7. http://www.defense.gov/releases/release.aspx?releaseid=11335.
8. "What is Galileo?". ESA. Oct. 2011 http://www.esa.int/esaNA/GGGMX650NDC_galileo_0.html.
9. http://www.cnbc.com/id/38738810/Spying_For_Profits_The_Satellite_Image_Indicator?slide=4 http://rsmetrics.com/.
10. Manyika, J., Chui, M., Brown, B., Bughin, J., Dobbs, R., Roxburgh, C., Byers, A.H., "Big Data: The Next Frontier for Innovation, Competition, and Productivity", McKinsey Global Institute, May 2011.
11. Gurtuna, O., Space for Energy: The Role of Space-based Capabilities for Managing Energy Resources on Earth, in P. Olla (ed.), "Space Technologies for the Benefit of Human Society and Planet Earth", 2009.
12. Gurtuna, O., and Prevot, A., "An Overview of Solar Resource Assessment Using Meteorological Satellite Data," in proceedings of the Recent Advances in Space Technologies Conference, Istanbul, Turkey, 2011.

Chapter 3
Seven Distinguishing Features of Space Business

Even though the fundamental rules of economics apply in all industries, there are several features of the space industry that set it apart.

Cyclical Nature

All of the three key indicators of economic activity discussed earlier (private sector revenues, government spending on space and employment figures) fluctuate over time based on the changes in demand and supply for space-related goods and services. The "boom and bust" periods are by no means unique. Almost all other economics sectors are prone to significant changes over time.[1] However, the long investment horizon of space activities and the long lead times associated with the design, manufacture and launch of space assets, which we will shortly discuss, exacerbates the problem.

Linkage to Defense

Since the very early days of the Space Age, space activities have been inextricably linked to defense priorities. The increasing prominence of commercial space activities has weakened this dependence a bit. Nevertheless, just like any other technology, space technologies can be used for both civilian and military purposes. This "dual use" nature of space technologies and applications is in the "genes" of the space market. The link to defense is not just limited to the core technologies of the space industry, such as launch vehicles, high-resolution optical imagery and high-bandwidth telecommunications. Other important commonalities include project management practices, export restrictions, recurrent cost over-runs and dependence on government budgets.

Many of the technological achievements in the space sector can be traced back to military objectives. Intercontinental ballistic missiles (ICBMs) evolved into reliable launch vehicles, the need for the positioning and navigation of military forces spurred the Global Positioning System, the electro-optical technology of spy satellites paved the way for many civilian applications that depend on high-resolution satellite images. Thus, it is not surprising that almost all of the key defense and space contractors are active in both the military and civilian markets.

Being active in both of these market segments enables companies to diversify their client base and hedge against deep spending cuts in government budgets or collapsing private sector demand when there is an economic recession. It also facilitates the diffusion of key technologies and capabilities gained through defense contracts into commercial applications. However, the dual-use nature of space business also imposes many restrictions on how and where business can be conducted. In this regard, the classic example is "International Traffic in Arms Regulations" (ITAR), a set of U. S. government regulations that control the export and import of defense-related goods and services. U. S. lawmakers have included many key space industry components in the ITAR list, which severely limits the capability of U. S. space companies to deliver service to international clients. As a result, Intelsat, Inmarsat, and other satellite operators are increasingly choosing suppliers from Europe, Canada, Australia, Japan, Russia and China to minimize ITAR-related risks.

This limitation not only covers the hardware and software that goes into a satellite, but it also restricts the launch vehicles that can be used to place the satellite into its designated orbit (Chinese launch vehicles, for example, are definitely off limits). Imposing such restrictions on a global marketplace naturally creates problems for U. S. companies. It is estimated that the opportunity cost for the U. S. businesses was around $2.35 billion for the period of 2003–2006.[2]

Interestingly enough, even non-U. S. companies have to be vigilant about ITAR, as the U. S. State Department can start investigations if they believe that a U. S.-made component is included in a satellite manufactured by a non-U. S. entity.[3]

Government as the Main Customer

Today, the number one client of space-related products and services is still the government. Even in well established segments such as satellite telecommunications, a sizeable part of the market is comprised of "dual use" services sold to governmental entities.

The active involvement of the government in the space sector has been a constant since the beginning of the Space Age – first as a technology provider, then as the main customer and regulator, and today as the anchor tenant for private space exploration. The bitter rivalry between the United States and the former USSR during the Cold War resulted in a substantial public investment in space activities on both sides of the Atlantic. During the Apollo program, the United States spent nearly 0.8 % of its Gross Domestic Product (GDP) for funding NASA's activities.

In comparison, today the United States spends about 0.3 % of its GDP on funding its civilian and military space activities, according to OECD's official estimates.

The main themes of the race between the United States and the USSR evolved over time as major milestones were reached one by one. In the 1950s, the competition was to develop improved launch vehicles. In the 1960s, the efforts on both sides of the Iron Curtain were focused on improving human spaceflight capabilities, with the ultimate target of landing on the Moon. By the 1970s, space stations were the main theme, as the Salyut-series space stations of the USSR were pitted against the SkyLab of the United States.[4] During these three decades, the driving force behind the world's major space programs was highly political, and a sound economic rationale was not strictly necessary to justify investments. The main role of the private sector during this period – especially in the U. S. space program – was that of a contractor: meeting the needs of the government within budget and on schedule. Even the advances in the satellite telecommunications domain were taken largely at the initiative of governments with the establishment of intergovernmental entities, such as Intelsat (a U. S.-led initiative) and the InterSputnik (led by the USSR).

It is important to note that the motivations of the private sector and the public sector can be – and usually are – fundamentally different. A profit maximizing firm will seek to establish an efficient, cost-effective business model targeting the lucrative segments of the market. A government, on the other hand, may be more interested in building industrial capability, perhaps fostering international cooperation or frequently seeking to provide universal service covering all of the market segments whether they are profitable or not. A good example is the postal service. Without a universal service mandate, most postal organizations would under-serve the rural markets and concentrate on larger urban areas.

Heavy government involvement in an industry such as space is not necessarily a disadvantage. In fact, when the economy goes into a recession, the stability of government contracts can act as a counter balance and help companies weather the storm. But this stability comes with a cost. In the space industry, very few companies bet on risky but rewarding new projects without any help from the government. However, as discussed in Chap. 5, a new generation of entrepreneurs is determined to shake things up and attract more private capital into the space industry.

The Destination Problem

One of the main issues with space transportation is that, in most cases, there is very little need for serving destinations in space on a regular basis. Currently, the only destination served regularly is the International Space Station. Even then, there is currently a single human-rated transportation system (the Soyuz) and a handful of cargo vehicles available. The location of the end user largely determines the nature of the transportation. Currently, with the exception of the ISS, all space products and services are destined to serve end users on Earth. Suborbital space tourism may increase the demand for launch services, but since most of these flights will start and end at the same location, their appeal to businesses will be limited. Point-to-point suborbital flights may partially surmount this problem. The early efforts in suborbital space tourism may morph into hypersonic transportation systems, dramatically reducing the transit times for long-distance air travel.[5]

In the long-term, if new sources of demand start appearing on other planetary bodies, the need for connecting the demand and supply nodes will naturally increase. Without a clear commercial need to establish sustained robotic and/or human presence on the Moon, Mars and other planetary bodies, the case for high volume production of launch vehicles will be limited.

Limited Competition

Competition stimulates innovation. The forces of "creative destruction" aptly described by Joseph Schumpeter play a critical role in any given industry by constantly replacing the established practices and technologies by a new generation of innovations.[6] The intensity of the competition largely determines the pace of this renewal process. The explosion of technologies and new business models related to the Internet is a case in point. By comparison, up until very recently, the composition of the space industry was very static. This lack of innovation can be attributed to the dominance of large integrators who control the supply chain and the high barriers to entry. The latter is mostly a result of the very large amounts of investments needed to meet the non-recurring costs (e.g., assembly and testing facilities, launch pads, etc.). The emergence of SpaceX, Virgin Galactic, Stratolaunch Systems, XCOR, and many other newcomers has slowly started changing this picture.

Long Investment Horizon

Another key factor that sets the space industry apart is the very long lead-times associated with most space products and services. From blueprint to launch pad, it takes many years to bring an idea to orbit, even though there may not be any schedule slippages (although delays are quite common). There are numerous reasons for these long lead-times – the technical complexity of space projects, the frequent need to build complicated facilities for manufacturing or testing processes, time-consuming space qualification tests, regulatory overhead (e.g., spectrum allocation, ITAR clearance, etc.) and transit times for planetary missions.

A government may invest in an infrastructure project, such as a new launch site or a new generation of launch vehicles, and wait for many years for it to be built and then generate socio-economic benefits. Companies, on the other hand, have much shorter investment horizons, and they are under tremendous pressure from the shareholders to deliver profits as soon as possible.

The Curse of the Single Unit of Production

Economies of scale is one of the most powerful forces in economics. Decreasing the unit cost of production unleashes the potential of mass-market adoption of goods and services. In the absence of economies of scale, unit production costs are very unlikely to go down, since the one-time R&D, engineering, design, and commercialization costs can only be divided in a limited number of ways.[7]

When an aerospace company, such as Boeing or EADS, designs a new aircraft, it is essentially creating a platform that will enter mass production. Income streams will be generated from the sale of the actual aircraft as well as from multi-year servicing and training contracts. In the space industry, this model is an exception and not the norm. Although there have been various attempts at mass producing satellites, such as the Iridium, Globalstar, Orbcomm and Teledesic series, this feat has to be repeated across the value chain (including launch services) to have a meaningful effect.

Case Study: Iridium and the Lessons Learned from Terrestrial Competition

The old saying of "If you want to make a million dollars in the space business, start with a billion" was right on the mark for the case of the Iridium LEO constellation. In the early 1990s, a wave of optimism and bold ideas managed to attract private capital to build mobile telecommunications systems that were to span the whole globe. Iridium was not alone in this journey, many other systems also managed to attract investors, and the order books of satellite manufacturers ballooned with projects from Iridium, Globalstar, ICO and other ventures.

After many years of design, development, manufacturing and a very successful launch campaign, Iridium started operations in October 1998 with 66 satellites in LEO. The total cost of building the system was in the range of $4–5 billion. Iridium's business model was essentially to be a mobile network operator in space. The intention was to sell handset units and establish a large subscriber base of users that would access the system for voice and data with the convenience of global coverage. Iridium executives made a bet on the premise – confirmed by several respected consulting firms – that a vast global market for mobile telephony was about to emerge. They were absolutely right about this premise. Today, there are billions of mobile connections around the world,[8] but only about 1.5 million of these are satellite-based.

For Iridium, one very serious challenge right from the start was finding enough paying customers to recoup the initial investment. The original plan was to attract half a million customers within the first 6 months of operations, and then expanding the customer base to 32 million subscribers within a decade. To this end, Iridium spent $180 million on advertising alone. Sadly, just 9 months after it began operations, the firm filed for bankruptcy, having acquired only 15,000 subscribers.[9]

In the year 2000, a consortium of private investors purchased the assets of Iridium for a mere $25 million. The U. S. Department of Defense jumped in as an "anchor tenant" and agreed to pay $3 million per month to provide voice and data services for 20,000 of its personnel. It's amazing how successful a business can be, if one can simply write off an initial investment of $5 billion and start from scratch. After many turns of corporate restructuring and changing hands of ownership, Iridium is still in business today, with a subscriber base of 500,000 customers and annual profits of $40 million (for 2011).[10]

With the luxury of hindsight, there are three main lessons we can learn from this multi-billion dollar exercise:

1. Cost is a significant barrier for space ventures. Access to space remains very expensive, and unless there are significant cost reductions in launch services, building space ventures from scratch will continue to require millions of dollars. Thus, developing business models that are based on leasing capacity from existing space infrastructure should be investigated first before any investment is made for the space segment. The Iridium experience demonstrated that even with economies of scale (e.g., high number of production units, shared launches, etc.) space business is extremely challenging.
2. Long lead times for space projects can be very dangerous, especially in fast-changing markets such as telecommunications. By the time Iridium had been deployed, the terrestrial competition (mobile networks) had already transformed the market with a robust product offering at a significantly lower cost. When Iridium finally entered the market in 1998, most of its potential customers were happily using ordinary mobile phones that were smaller, lighter and offered much better quality of service than Iridium handsets.
3. Determining the evolving needs and wants of the customer is very difficult. Although global coverage is an impressive feature, it turns out that the primary customer segment of Iridium, the corporate market, has no particular need to make phone calls in the middle of a desert or ocean. Also, the call quality of the handsets was patchy at best when used indoors, a significant problem for office workers. However, certain other customers do need global coverage in the great outdoors in rural and remote areas. This more targeted clientele included such users as the military, oil and gas, and mining industries. Once Iridium was able to cater to these customers, it was able to expand its customer base (although its sales were never anywhere close to the original projections).

Notes

1. For building insight regarding business cycles, see the excellent interactive simulation "Beer Game" http://web.mit.edu/jsterman/www/SDG/beergame.html.
2. Tkatchova, S "Space-Based Technologies and Commercialized Development: Economic Implications and Benefits," Idea Group Inc, 2011.

Notes

3. De Selding, P. "U.S., Thales at Odds Over Request for ITAR-free Satellite Design", Space News, Jan 9, 2012. http://advertise.spacenews.com/sn_pdfs/SPN_20120109_Jan_2012.PDF.
4. For an in-depth account of the policy rationale behind these investments, please see Logsdon, J. M., "John F. Kennedy and the Race to the Moon", Palgrave Macmillan, New York, 2010.
5. FAA, *"Point-to-Point Commercial Space Transportation in National Aviation System: Final Report"*, March 10, 2010, Washington D.C., available at http://www.faa.gov/about/office_org/headquarters_offices/ast/media/point_to_point.pdf.
6. Schumpeter, J. A., "Capitalism, Socialism and Democracy," Harper Perennial, 1962 (reprint of the 1947 edition).
7. Some experts believe that we may be at the threshold of a manufacturing revolution which may enable producing at lower volumes in a cost-effective way (see for example The Economist, "The Third Industrial Revolution", April 2012). The convergence of software, novel materials, robots, and new processes (such as three-dimensional printing) may have profound effects on many industries, including space.
8. Gartner, "Worldwide Mobile Connections Will Reach 5.6 Billion in 2011 as Mobile Data Services Revenue Totals $314.7 Billion", Press Release, Stamford, Conn., August 4, 2011 available at http://www.gartner.com/it/page.jsp?id=1759714.
9. Economist, "Iridium, A New Orbit: The Ill-Fated Satellite-Phone Venture Has Relaunched Itself In More Modest Form", The Economist, London, Jul 12th, 2001.
10. Iridium, "Fourth-Quarter and Full-Year 2011 Results", Press Release, March 6, 2012 available at http://investor.iridium.com/releasedetail.cfm?ReleaseID=654525. Iridium Announces.

Chapter 4
Socio-Economic Benefits of Space Activities

In 1924, when asked why he was willing to risk his life to climb Mount Everest, British explorer George Mallory had a very simple answer: "Because, it's there." This iconic phrase is now synonymous with our profound desire to explore. Although a similar answer could have been acceptable for space exploration in the 1960s, justifying the investments in space activities require a much more utilitarian rationale today.

So, why do governments invest in space activities? Answering this question was relatively easy at the beginning of the Space Age. Space was a high-tech arena where rival ideologies clashed, showcasing their technological prowess under the watchful eyes of the entire globe. The competition between the former USSR and the United States brought about a rapid succession of firsts – the first human in orbit, the first steps on the Moon, the first space station and the first reusable launch system. However, once the dust settled, and the eye-popping bill of the early space activities hit government coffers, a new rationale emerged: pragmatic use of space as the higher ground for improving life on Earth.

From satellite telecommunications to remote sensing, from microgravity research to satellite navigation, government space programs started investing in practical outcomes for the public good. Therefore, measuring the socio-economic benefits of space activities steadily gained importance throughout the 1970s and 1980s, an era defined as "Mission to Planet Earth."

Measuring the benefits of space investments is no easy feat. As discussed earlier, the benefits of space activities accrue over a long period of time. In some cases this can be decades after an initial investment is made. Furthermore, tallying up the benefits is generally not easy, as a large portion of the purported "gain" can be intangible and based on societal benefits rather than private sector revenues.

Three Main Types of Benefits

Given the high profile of space in the public agenda, economists have devoted considerable attention to measuring the economic impacts of space programs. The economic effects of space programs can be broken down into three main components – direct benefits, spin-offs and indirect societal benefits.

Direct Industrial Benefits

These benefits are generated from the sales of products and services that are directly linked to the original purpose of a space project (e.g., the revenues from the sales of high resolution optical imagery either to a private sector or public sector client). Since the 1950s, space programs have generated a wide range of hardware, software and processes that have made their way into a myriad of applications. The benefits of these applications are directly attributable to the original investments made by the space agencies and the private sector. Although we are fairly certain of the amount of investments that went into these projects, measuring their direct economic impact is only possible if we can track their corresponding economic output over time and understand the nature of the contribution.

Spin-offs

In addition to the economic outputs directly attributable to space programs, there are also many "positive externalities." These are benefits generated by the space programs without prior planning. A classic example is a "spin-off": a technology or capability originally developed for space programs that was transferred to other economic sectors over time.

One distinguishing feature of spin-offs is their serendipitous nature. In virtually all cases, the indirect economic benefits that stem from spin-offs were not specifically anticipated or planned during the initial investment. These benefits thus tend to appear over time, as the technologies, best practices and tacit knowledge originating from space investments make their way to other sectors of economic activity.

Although the most visible type of spin-off is technology transfer, this is by no means the only kind of spin-off. Most space projects require the development of new bodies of knowledge, including best practices and management techniques. Over the years, space agencies and contractors generated new ideas, products, and new organizational methods as they gained more expertise. This learning process then diffused to other departments of large aerospace contractors and subsequently throughout the economic system itself. For example, Technology Readiness Levels, a widely used technology assessment tool for space projects, is not a technology by itself, but it's a very useful method that is now used in many different sectors.

Measuring the economic impact of spin-offs is a meticulous task that requires extensive interviews with space and non-space companies. Studies performed in Europe indicate that the estimated economic benefits accrued from spin-offs are about three times the original government investments.[1]

Societal and Intangible Benefits

The third significant type of benefit is not confined to a particular company or to the transfer of a specific technology. Rather, it is based on the positive impact of a space project on the society as a whole.

For instance, meteorological satellites have significantly improved the accuracy of weather forecasting, which in turn has improved our quality of life through incremental improvements in many economic sectors (e.g., tourism, maritime operations, airlines, insurance, etc.). Thus, a certain portion of these improvements can be attributed to the original investment.[2] Making the case for the societal benefits of space applications is relatively straightforward. All space programs dedicated to satellite applications have created some form of direct societal impact with associated economic benefits. These benefits can be measured using traditional cost-benefit analyses by identifying the net benefit for each economic sector.

Interestingly, history may prove that some of the most important benefits of space are the intangible ones. Past space missions have dramatically altered the way we perceive ourselves as a species. In 1996, Hubble Space Telescope produced an amazing image called the Hubble Deep Field. The director of the Space Telescope Science Institute, Robert Williams, decided to use his discretionary observation time for the study of distant galaxies. This, of course, is not unusual at all. What was extraordinary was the intense focus of Hubble for this observation. For ten consecutive days the telescope was pointed towards a tiny speck in the sky. The resultant image revealed at least 1,500 galaxies. The sheer number of galaxies in this image is one of the visual landmarks of space exploration. Although we know how much the Hubble Space Telescope cost (about $12 billion over the last 30 years, including the cost of five shuttle servicing missions), it is almost impossible to attach a monetary figure to knowing our place in the universe (Fig. 4.1).

The table below summarizes the type of socio-economic benefits that accrue from space investments (Table 4.1).

Measuring the Economic Impacts of Space Programs

As discussed in Chap. 2, the global economic footprint of the space industry, including both public and private sectors, is in the range of $235–277 billion. Although these figures may look impressive at first sight, the space industry is actually quite small compared to the other sectors of economic activity. The Gross World Product (GWP) is the total gross national product of all the countries in the world (based on

Fig. 4.1 Hubble Deep Field (Source: R. Williams and the HDF Team and NASA)

Table 4.1 Overview of socio-economic benefits

Type	Category	Summary of benefits	Examples
Direct industrial benefits	Space as an enabler	These benefits can directly be attributed to the corresponding space technology. Without this technology, the benefits simply would not exist	Global navigation; direct-to-home television
	Space as a contributor	Certain space-based capabilities improve our quality of life and also help reduce the cost of our daily activities. Space may not play the central role, but nevertheless it provides an important contribution	Forecast of extreme weather events (e.g., hurricanes); tracking icebergs and sea-ice; precision farming using satellite navigation
Spin-offs		These are benefits that accrue from adapting a space technology to a terrestrial domain. These benefits may or may not have been foreseen as part of the technology development effort	Contrary to popular belief, Tang, Velcro, and Teflon are not actually in this group, but memory foam and robotic surgery are
Societal and intangible benefits		These are benefits that cannot necessarily be quantified in the economic sense as they relate to intangible achievements, such as public awareness, prestige, political gains and international recognition	Hubble Deep Field; national prestige (especially for human spaceflight); understanding the nature of climate change and the extent of the challenge to human civilization

purchasing power parity). For 2011, the GWP is estimated at $79 trillion.[3] Thus, the entire space economy does not represent even 1 % of the global economic activity.

At the beginning of the Space Age, with all sights firmly set on prestigious prizes, such as being the first in orbit and then first on the Moon, the economic rationale of space investments was not a priority. However, as these milestones were reached over time, the political momentum started to fade. The cost of the Apollo program was the primary reason for its cancellation in 1972. Following the lunar landings in 1969 and the early 1970s, the Nixon administration was faced with the difficult problem of how to "get NASA's budget under control" while still maintaining the lead of the United States in the space race. Sending more American astronauts to the Moon would not have brought more prestige, and without an economic basis for sustained flights to the Moon, there was very little reason to keep the Apollo program running.[4]

Once the excitement of the 1960s subsided, measuring the economic benefits of space programs became more and more important. In the absence of a strong political mandate, the economic rationale started to shape major investment decisions. The initial attempts to quantify the economic impacts mostly centered on the macroeconomic picture. By using econometric models, researchers tried to identify and measure the portion of economic growth attributable to space activities. One of the most cited studies was performed by the Midwest Research Institute (MRI). The MRI study was contracted by NASA and looked into the relationship between NASA R&D expenditures and technology-induced increases in the U. S. Gross National Product (GNP). This study concluded that each dollar spent by NASA on R&D during the 1950–1969 period returned an average of slightly over $7.00 in GNP through 1987.[5]

Since this study, the 1 to 7 ratio of R&D investments to economic returns has been widely used as a way to justify investing in space. Although the early days of the Space Age generated a very significant economic return, there are some inherent dangers in blindly using this ratio today:

- The ratio is an average figure. Some R&D investments had generated many times their original public investment while some other investments had negligible returns. Thus, without looking into the specific benefits expected from a space investment, there is no guarantee that the returns will be in a similar range.
- The marginal returns of space investments have decreased over time as many technical challenges were surmounted by innovative products and services. This is not to say that space investments in the future will not generate significant returns, but it is only natural that the initial investments unlocked more value than subsequent ones.
- Some of the accrued benefits are societal in nature (such as gaining a better understanding about climate change, or a heightened sense of planetary protection as we learn more about the past of Mars and Venus). Such gains may not have a direct economic benefit (at least in the short term), but they have contributed greatly to our collective knowledge of nature.

Most of the studies we mentioned conclude that space activities have created significant economic value for the whole economy through the creation of new products and services, transfer of new technologies and many positive externalities, such as social and environmental consciousness.

A Deeper Look at Spin-offs

The infancy of the space industry resulted in a very suitable environment for cross-fertilization of ideas and transfer of many technologies from space to other sectors. The time between the 1950s to the mid-1980s was largely defined by spin-offs during which space technologies made their way to many other applications. Through time, it's only natural that the quality and quantity of the economic effects will change. For example, if the spin-offs were an important side benefit of space activities up until the mid-1980s, as time passes, there may be less emphasis on spin-offs and more on "spin-ins" (e.g., the diffusion of technologies from terrestrial sectors into the space industry).

Technology transfer is not a very orderly process, and it's certainly not unidirectional. It is interesting to note that after decades of extensive spin-offs, we are now seeing more two-way traffic between space and terrestrial domains. In fact, we should be expecting more spin-ins in the coming decades and the convergence of multiple technology areas that can have a tremendous impact on the future of space exploration.

This trend towards decreasing intensity of spin-offs is a result of several factors, including regulatory mechanisms, such as ITAR, which are specifically designed to hinder technology transfer due to national security concerns and the longer lead times associated with the design, manufacturing and deployment of space assets.

Nevertheless, spin-offs can be an excellent point of departure for entrepreneurs who are interested in developing new commercial applications. Space agencies actively promote the transfer of space technologies to other economic sectors; in some cases, free licenses can be obtained to gain access to space technology.

The European Space Agency's Business Incubation Centers provide very interesting information regarding a wide-range of space technologies that were successfully applied to terrestrial domains. An analysis performed in 2011 clearly demonstrates the depth and breadth of the dissemination of space technologies.[6] Lifestyle, software solutions, environment and health are some of the main sectors that have benefitted from space technologies. The reach of space technologies extend to many other sectors as well, including energy, textile, automotive and life sciences.

It is also interesting to track the origin of the ESA's spin-off technologies, spanning the period of 1990–2006. During this period, space science and launchers were the two leading domains of space technology, accounting for about 20 % of the spin-offs each. Human spaceflight, microgravity research, telecommunications and Earth observation contributed to around 10 % of the spin-offs each.

Adapting space technologies to meet different needs on Earth can unlock tremendous value. The table below shows the link between space technologies and applications in medicine, manufacturing, entertainment and many other sectors (Table 4.2).

Table 4.2 From space to Earth: spin-off examples[a]

Space program technology and commercial spin-offs	
Product	Space origin
Tumor tomography	NASA scanner for testing
Battery-powered surgical instruments	Apollo Moon program
Non-reflective coating on personal computer screens	Gemini spacecraft window coating
Emergency blankets (survival/anti-shock)	Satellite thermal insulation
Mammogram screening, plant photon-counting technology	Space telescope instruments
Skin cancer detection	ROSAT X-ray detection
Dental orthodontic spring	Space shape memory alloys
Early detection of cancerous cells	Microwave spectroscopy
Carbon composite car brakes	Solid rocket engine nozzles
Car assembly robots	Space robotics
Flameproof textiles, railway scheduling, fuel tank insulation	Various Ariane components, including software
Lightweight car frames, computer game controllers, fuel cell vehicles, coatings for clearer plastics, heart assist pump, non-skid road paint	Various space shuttle components
Fresh water systems	ISS technology
Corrosion free coating for statues	Launch pad protective coating
Flexible ski boots, light allergy protection, firefighter suits, golf shoes with inner liner	Various spacesuit designs
Healthy snacks	Space food

[a]The following sources were used to compile the table: Peeters, W., "Space Economics And Geopolitics", ISU Executive Space MBA lecture notes, 2001; ESA, "Down to Earth: How Space Technology Improves Our Lives", 2009 available at http://esamultimedia.esa.int/multimedia/publications/BR-280/pageflip.html and NASA Spinoff website available at http://spinoff.nasa.gov/

Case Study: On Stardust and Dollars

In February 1999, NASA launched a mission named "Stardust" with the primary aim of collecting samples from a comet and returning them to Earth. After collecting cometary particles and interstellar dust, mission controllers successfully landed a return capsule in the Utah desert in 2006. In total, Stardust collected about 100 interstellar dust particles and a few thousand cometary particles.[7]

It is relatively easy to estimate the cost of these samples to the taxpayer. The reported budget of this mission was $208 million,[8] so each sample cost the taxpayers anywhere from $100,000 to $1 million, depending on the exact number of the samples. However, the real challenge is to estimate the socio-economic benefits of these samples. In other words, what is the value of a speck of stardust?

Applying the traditional cost benefit analysis tools, such as net present value, would give misleading results, since the economic benefits have to be expressed in strictly monetary terms. In the case of the Stardust mission, the science team discovered that the traditional definition of comets as clouds of ice, dust and gases is not very accurate, as there were remnants of high-temperature materials in the samples. Another surprise was the origin of the comet. The science team expected to find particles belonging to other solar systems much older than ours. When the samples were analyzed it was found that the comet was formed in our very own Solar System. These findings have definitely caused many textbook chapters to be rewritten, and they also provided an important step in truly understanding the origins and dynamics of our Solar System. There was also a bonus – the aerogel developed for this mission to collect the sample particles was listed in the Guinness Book of World Records as the lightest known solid material. Maybe in a couple of decades many economic benefits will accrue from the use of this material in our everyday lives.

Notes

1. Bach, L., Cohendet, P., Schenk, E., "Technological Transfers from the European Space Programs: A Dynamic View And Comparison With Other R&D Projects", The Journal of Technology Transfer. 27(4):321–38, 2002.
2. Although tallying up the societal benefits can be fiendishly difficult. For an overview of these valuation challenges, see Macauley, M. K., "Investing in Information to Respond to a Changing Climate", Resources, Summer 2011, available at http://www.rff.org/documents/about_rff/donor_events/120112_dc_megacity_carbon_project/rff-resources-178_feature-macauley.pdf.
3. CIA, The World Factbook, World Economy, available at https://www.cia.gov/library/publications/the-world-factbook/geos/xx.html.
4. In fact, a loss of life during an Apollo mission (after the first Moon landing) might have tarnished the reputation of the United States as a technological powerhouse.
5. The MRI study estimated NASA's R&D spending during the 1959–1969 period at US $25 billion (in 1958 dollars). The corresponding return on this investment was estimated at US$181 billion between 1959 and 1987.
6. Szalai, B., "A Quantification of Benefits Generated by ESA Spin-offs", International Space University Working Paper, 2011.
7. Finding Stardust, Stardust@Home, available at http://stardustathome.ssl.berkeley.edu/ss_findingsd.php.
8. NASA press release, "Missions To The Moon, Sun, Venus And A Comet Picked For Discovery", February 1995, available at http://nssdc.gsfc.nasa.gov/planetary/discover95.txt.

Chapter 5
Emerging Space Markets

The turn of the new millennium has not been very kind to the space industry. Ambitious mobile telecommunications systems financed by the private sector in the 1990s either went bankrupt or never left the launch pad. Two economic recessions that have occurred within the last decade put even more pressure on the anemic growth in space agency budgets. Finally, the retirement of the space shuttle in 2011 left the United States without an indigenous human spaceflight capability to orbit for the first time since 1962.

These events mark an era of deceleration in space activities. However, as we discussed earlier, just like any other economic sector, the space industry also has a cyclical side. Tides do turn eventually, and powered by a number of trends, the space industry seems to be at the beginning of a new wave of growth. One of these trends is the impact of globalization, a second one is the rise of the private sector and a third one is the ascendance of emerging markets, the topic of this chapter.

We can define emerging space markets as markets that are not significant sources of economic activity today, but those with a big potential for growth. The growth of emerging markets can energize the global space industry with new waves of infrastructure investments, innovative product and service offerings, and a healthy dose of entrepreneurial dynamism.

Emerging Sectors

Without any doubt, existing markets, such as satellite telecommunications, will continue to be important sources of new services and products in the industry, especially as more countries continue along their path of economic development and require the use of space-based capabilities. However, we can also expect significant momentum in entirely new space-based markets in the future. Although there are many candidates, for sake of brevity we will focus on three main emerging markets in this chapter: space tourism, on-orbit satellite servicing, and private space exploration.

Space Tourism

To date, just over 500 humans have ventured out into space. Proponents of space tourism aim to increase this number drastically. According to the World Travel and Tourism Council, the total economic value of goods and services attributable to tourism in 2011 was around US$ 5.98 trillion,[1] or 7.5 % of the Gross World Product. Compared to the economic footprint of the space industry, this is an immense market, and if the space industry can develop products and services that target the tourism sector, significant growth can be achieved in commercial space activities.

The space tourism market has three main segments: terrestrial/high-altitude, suborbital and orbital. The terrestrial/high-altitude segment covers a wide range of services, such as visits to space centers, ground simulators, parabolic flights, and flights in fighter jets. The suborbital segment is comprised of short-duration suborbital flights up to an altitude of 100 km. Finally, the orbital segment includes visits to LEO and beyond.

Although there is an enormous difference between these segments in terms of the complexity of the offerings and the sheer energy requirements to reach different altitudes, the public perception generally lumps them together. For example, although there is only a couple hundred kilometers of difference between the altitude of a suborbital flight versus that of an orbital flight, reaching orbit requires about 25 times more energy. Furthermore, getting up there is only half the challenge. Returning safely back to the surface of Earth requires disposing of most of this energy.[2] As the *Columbia* accident painfully reminded us in 2003, this is not a trivial task.

By any account, the space tourism market is still in its infancy, although there have been many remarkable achievements within the last decade. Starting with Dennis Tito in 2001, a total of seven paying customers visited the International Space Station. One of them, Charles Simonyi, completed two trips. In 2004, a U. S. company, Scaled Composites, headed by aerospace guru Burt Rutan, won the Ansari X-Prize by successfully reaching the suborbital attitude of 100 km twice within a 2-week period. This technical feat was financially backed by Microsoft co-founder Paul Allen. Following these successful flights, the spacecraft, SpaceShipOne, and the carrier vehicle, White Knight, are now being commercialized with the launch of a new enterprise known as Virgin Galactic. This undertaking, a venture of serial entrepreneur Sir Richard Branson, has now evolved into a leading contender in the emerging suborbital space tourism market.

Currently, the newly designed fleet of Virgin Galactic is being built by Scaled Composites, with a view to start commercial operations as early as 2013. According to Sir Branson, as of May 2012 the company had already collected $200,000 each in prepayments from 550 customers.[3]

Since 1994, more than ten separate market studies have been conducted to assess the revenue potential of the suborbital segment. Most of these studies indicate that there is a strong demand for space tourism. Price elasticity of demand, a concept explained earlier, is very useful in gauging the level of demand at various price points. All things being equal, as the ticket prices go down, more people will be able to afford suborbital flights. The sweet spot of the market seems to be around $50,000 per ticket, which yields the maximum revenue per year of around $785 million, from nearly 16,000 passengers.[4] Whether or not this volume of traffic can be safely supported is an entirely different question.

Despite all the marketing hype behind space tourism ventures, long-term success is critically dependent on the safety of suborbital spacecraft. Multiple decades of R&D, testing and operational experience enabled passenger aircraft to attain a remarkable degree of safety. There will always be thrill-seeking individuals who may be convinced to hop on the next suborbital vehicle right after a catastrophic accident, but for the general public to start flying regularly, many decades of routine operations will be needed.

If space tourism remains as a luxury market segment that caters to customers seeking adrenaline and boasting rights, its appeal is going to be exclusive but limited. It will also attract very few repeat customers, since the cost and risk of taking additional flights will probably not be acceptable for many initial customers. In this regard, the space tourism market may also suffer from the destination problem we alluded to in Chap. 3.

If, on the other hand, it can evolve into a necessity for conducting business, even an accident may not deter its continued growth. One way out of this conundrum may be to develop services such as point-to-point transportation. By connecting major cities around the world with suborbital flights, very significant reductions in travel durations may be achieved. For example, it takes about 13 hours for a conventional airliner to fly from New York to Tokyo. A suborbital vehicle on a ballistic trajectory might complete the same flight within 1.5–2 hours (although gate to gate flight time will be longer due to boarding and deplaning procedures).[5]

Before orbital space tourism can evolve into a significant market, there are a number of necessary conditions that have to be met. One major challenge is on the supply side – the lack of reliable and frequent access to LEO and beyond. Companies such as SpaceX and Scaled Composites have embarked on ambitious programs to develop the next generation spacecraft. Their success in this endeavor will largely determine the future of this market segment.

However, there are also many challenges on the economic front. The space tourism industry cannot solely count on millionaires as customers: sooner or later it has to be able to attract the general public. Lowering the price point may be possible with economies of scale and efficient operations. This feat may be achieved in the suborbital segment decades before the orbital one.

On-Orbit Satellite Servicing

The conventional business model for any space-based venture is severely limited by the amount of supplies that can be carried as part of the mission. Once the vital supplies of a spacecraft are depleted (such as the on-board fuel for station keeping) and/or a critical subsystem fails, the spacecraft can no longer function normally.

The painstaking process of resupplying the International Space Station with many different consumables, spare parts and fuel is a case in point.

One way around this limitation is to deploy robotic systems that are capable of maintaining and/or repairing a spacecraft. Called "on-orbit satellite servicing" (OOS), such a capability can drastically increase mission lifetimes, change our current way of managing risk in space and increase the value of each spacecraft by adding operational flexibility. A historical analog to OOS is in-flight refueling, which has brought a number of benefits for military aviation missions, such as reduced mission cost and increased mission range and duration.

As we will see in Chap. 7, the current paradigm in risk management is to design spacecraft with multiple layers of redundancy. Although incremental improvements in the quality of components as well as maturing systems engineering practices significantly increased the expected service life of spacecraft in orbit, one remaining key limitation is the fuel used for on-board station keeping. Once this precious fuel is depleted, satellite operators lose their fight against the natural decay of the orbits and effectively lose control of their satellites. The other key "expendable" relates to batteries and their practical in-orbit life. Solar energy can recharge spacecraft batteries only to certain limits. Therefore, on-orbit refueling and/or battery replacement can be very valuable services as they can significantly increase the useful life of a satellite.[6]

The theoretical and technical foundations of OOS are already in place. In 1997, the Japanese experimental satellite mission, ETS VII, was used to verify rendezvous/docking and various robotics technologies for future missions.[7] In the United States, the Orbital Express program of the Defense Advanced Research Projects Agency and the establishment of the Satellite Servicing Capabilities Office at NASA created considerable momentum towards autonomous servicing, repair and refueling operations.[8]

The economic feasibility of OOS is a different question. There have been numerous attempts in the past to launch ventures to tap into this market. Companies such as Orbital Recovery Corporation and MacDonald Dettwiler and Associates Ltd. (MDA) have announced their plans to service GEO satellites.

One of the main challenges for OOS is the standardization of satellite interfaces, so that the client satellites can be captured and serviced. In this respect, modular designs for fuel tanks, standardized "spare parts" such as batteries and servicing protocols could be very useful in the further development of this market.

We will revisit on-orbit satellite servicing at the end of this book and use it as a case study to better understand the core concepts in space business and economics.

Private Space Exploration

Private space exploration is an umbrella term that covers a diverse set of activities. It includes sub-segments such as mining, launch services, and even space entertainment. Given all of the challenges of space business, one may be tempted to conclude that this is a very hostile environment for entrepreneurship. As we have discussed earlier, government is the key actor in the space sector, and most investment decisions are taken based on government priorities, especially when it comes to space exploration and space science activities.

A relatively recent development in the space industry is the emergence of the private sector as an end-to-end provider of space products and services to the general public or to other private sector entities. Known as "business to consumer" and "business to business" markets, respectively, these two types of transactions constitute the bulk of economic activity for many different industries. In the space market, satellite telecommunications and, to a certain extent, satellite navigation segments are closer to this model of economic activity.

Of course, the forces of entrepreneurship have always been present in the space market. Many companies were formed and many products were launched. However, up until now, in the space exploration domain, these companies exclusively catered to the government. Today, we may be witnessing a fundamental transformation. A growing number of determined and resourceful entrepreneurs have finally started disrupting the relatively static state of affairs.

These space entrepreneurs, also sometimes called "astropreneurs," have managed to convince private investors to invest in the space exploration market, despite the significant technical and financial risks involved. These investors share certain common traits – a fascination with space, and the know-how to launch successful ventures in high-tech industries, most notably in the IT sector. Space investors such as Paul Allen (of Microsoft fame), James Cameron (of Hollywood blockbuster fame), Jeff Bezos (founder of Amazon.com), Robert Bigelow (owner of Budget Suites), Sir Richard Branson (Virgin Enterprises entrepreneur), John Carmack (video game developer of such games as Doom), and Elon Musk (founder of PayPal and Tesla Motors), have all achieved remarkable successes in other sectors, and for most of them space provides the next big challenge for their careers. Since these prime movers in the commercial space sector are all billionaires they have a unique way of approaching capital financing that is not available to conventional start-up efforts. The involvement of billionaire space enthusiasts, however, is no guarantee of success. In 1990s, Bill Gates and Craig McCaw teamed up to back the Teledesic LEO satellite network that never made it past the design phase and went bankrupt.

Another clear distinction of space entrepreneurs is their attitude towards risk. They are prepared to take more risks than the space agencies, especially when it comes to human spaceflight. They also operate their organizations in a "lean" way, getting more done with fewer resources. Since the main objective of a private enterprise is not to fulfill societal objectives, such as creating employment, they have more freedom in allocating their resources. Thus, the work teams in these space ventures tend to be much smaller than their space agency counterparts, reducing the labor cost, a key driver of life cycle cost for space projects.

Contrary to the cold and deadly image of space, it's actually teeming with resources. In our own Solar System, in addition to an abundant amount of solar energy, planetary bodies harbor a host of resources, including water, silicon, helium 3, a variety of minerals, metals and other natural resources. Due to the harsh environmental conditions in space, exploiting these resources may not be feasible for humans in the short term. However, robotic precursors can achieve much more than proof-of-concept missions; they can actually prepare the ground for eventual human settlement. Our nearest celestial neighbor, the Moon, is rich in useful materials, especially around the lunar poles. In 2009, NASA's LCROSS mission provided scientific evidence for the presence of a substantial amount of water-ice in the permanently shadowed regions of the Moon.

The Ansari X-Prize changed many things in the space industry. Among these changes was the perception that the role of the private sector is mostly the one of a contractor. By designing, building and launching a suborbital spacecraft entirely by private financing, Scaled Composites not only showed that the suborbital challenge can be met, but it can also be turned into a full-fledged business operation with worldwide media interest.

The success of Scaled Composites in reaching suborbital altitudes inevitably raised the bar higher – reaching low-Earth orbit and beyond. One of the most anticipated new space ventures is Stratolaunch Systems, which brings together some of the top names in private space exploration. Paul Allen and Burt Rutan have partnered once again to build the world's largest aircraft as a high altitude rocket launcher. Instead of developing a launch vehicle on their own, they managed to convince Elon Musk's SpaceX to provide the rocket and Dynetics, a well-established aerospace firm, to provide the mating interface. With a wingspan of 117 m and six 747 aircraft engines, the carrier aircraft will inspire awe. But what is even more impressive is the partnership model: private sector entities grouping themselves without any government involvement. This may be the norm in many other industries, but when it comes to space exploration it's a novelty.

The rise of private space exploration is not a uniquely U. S. phenomenon. Other countries around the world have also started to explore this path. A very interesting example comes from Denmark in the form of Copenhagen Suborbitals. Solely based on private donations, this company managed to successfully conduct a suborbital launch in 2011 (Fig. 5.1).

There are many challenges ahead for private space exploration. Some of these are on the legal and regulatory side. Currently, there are various United Nations treaties that restrict the private ownership of planetary bodies. Despite what anyone might argue today, no one has the legal right to claim and sell "parcels of the Moon." However, in the near future, as companies start to establish a virtual presence on planetary bodies using robotic missions, the issue of responsible development of planetary resources by the private sector will be a key consideration.

Another key obstacle is the destination problem. Before we start analyzing the real potential of mining in space or building solar power satellites, we need to ask ourselves where the source of the demand actually lies. If the source of the demand is on Earth, there is a pretty good chance that a terrestrial substitute will be in place before the space-based service is operational (a distinct risk for space projects, as discussed in the Iridium case study). However, if the source of the demand is in space, then the business model can look fundamentally different. If, for instance, we could establish a materials-processing capability on the Moon with enough

Fig. 5.1 The successful launch of Copenhagen Suborbitals (Image credit: Bo Tornvig/Copenhagen Suborbitals)

sophistication that satellites could be manufactured on the lunar surface and then lowered to GEO orbit, this could allow for substantial savings in launch costs.

In the coming decades, the dynamism of the private sector can drastically alter the way business is done in the space industry. By assembling all the necessary components of successful ventures, including financing, technology, marketing and operational excellence, a new generation of companies can succeed where governments have largely failed. It is perhaps the ultimate challenge of private enterprise to create a strong economic rationale for investing in new and innovative space applications in orbit and beyond GEO.

Notes

1. WTTC, Travel and Tourism Economic Impact, 2011, http://www.wttc.org/site_media/uploads/downloads/world2_2.pdf.
2. Lindsey C.S., " Suborbital spaceflight: a road to orbit or a dead end?", The Space Review, December 15, 2003, available at http://www.thespacereview.com/article/73/1.
3. CNN interview transcripts, available at http://transcripts.cnn.com/TRANSCRIPTS/1205/22/cnr.07.html.

4. Futron, "Suborbital Space Tourism Demand Revisited", Whitepaper, Bethesda, MD, 2006 available at http://www.futron.com/upload/wysiwyg/Resources/Whitepapers/Suborbital_Space_Tourism_Revisited_0806.pdf.
5. ISU, "Great Expectations Assessing the Potential for Suborbital Transportation", Final Report, International Space University Masters Program, 2008.
6. ISU, "DOCTOR: Developing On-Orbit Servicing Concepts, Technology Options, and Roadmap", Final Report, International Space University, Summer Session Program 2007, Beijing, China available at http://isulibrary.isunet.edu/opac/doc_num.php?explnum_id=102.
7. Engineering Test Satellite #VII (ETS-VII/Orihime & Hikoboshi) Home Page, available at http://robotics.jaxa.jp/project/ets7-HP/index_e.html.
8. NASA, "On-Orbit Satellite Servicing Study Project Report", October 2010, http://ssco.gsfc.nasa.gov/images/NASA_Satellite%20Servicing_Project_Report_0511.pdf.

Chapter 6
Key Issues and Challenges in the Space Business

Improving the economic performance of existing space businesses or creating new ones both require thriving in a challenging environment. From the high cost of accessing space to regulatory hurdles, from the difficulties of finding investment capital to technical mishaps, there are many potential showstoppers. However, not all of these challenges are threats. Some of them can be turned into business opportunities through innovative business models.

High Cost of Access to Space

The fundamental technology behind launch vehicles has not changed since the beginning of the Space Age – chemical propulsion using solid and/or liquid propellants. One would assume that in the intervening five decades or so, the cost of this particular service would decrease significantly through incremental improvements in technology and economies of scale. Unfortunately, this has not been the case. Despite many years of research and development and operational expertise, launch costs are still very high. After the collapse of the Soviet Union, stockpiles of missiles were turned into launch vehicles, which caused a temporary decline in prices. However, once the stockpiles were used up, the average cost started going up again.

There are a number of initiatives that may bring the costs down. NASA's Commercial Orbital Transportation Services program was launched in 2006 in order to stimulate the efforts of the private sector to develop reliable and cost-effective means to deliver cargo to the orbit. SpaceX and Orbital Sciences Corporation are now actively working to develop cargo delivery services with a view to expand to human spaceflight in the future. However, even if these initiatives can bring a marginal decrease in the cost, we are still very far away from price points that can make transportation affordable enough to trigger a massive growth in demand for launch services (though it is not even a given that lower launch prices would spur higher demand, as discussed in Chap. 2).

Another key cost driver is the traditional business model of building a single flight unit that is customized to the needs of the end user. Although certain parts of the satellites can be based on standardized platforms (e.g., bus, solar panels, etc.), the frame of mind is still very much oriented toward building specialized products that are one off or just a few of a kind. Only the Teledesic initiative has sought to aggressively attack this problem head on to manufacture a thousand interchangeable satellites. In the words of system designer James Stewart, the idea was to "manufacture the Teledesic satellites like VCRs or TVs." This project, however, failed before the viability of this concept could be tested. The Iridium and Globalstar satellites with production runs of about 100 units moved closer to the mass production model, but with limited economies of scale.

In many other manufacturing sectors, such as aviation, automotive and industrial products, a high level of production volume is the norm. Whereas the current practice in the space sector is akin to designing and building a last generation airliner, producing a few units, flying them once and then moving on to the next design. This lack of economies of scale is one of the key reasons why the unit costs of space systems – for both satellites and launchers – are extremely high compared to other technology products. Only for some ground segment equipment (e.g., very small aperture terminals and microterminals) has major economies of scale been achieved, and hundreds of thousands of units were produced.

Limited Access to Financing

Another obstacle is the limited funding and types of investors who are interested in financing space projects. The lack of economies of scale, very long development cycles, technical risks and strict government regulations drive away most private investors. The mature segments of the space industry, such as satellite telecommunications, are able to access capital markets for their needs, but many other segments are almost entirely financed by the government.

Numerous attempts to use private-public-partnership (PPP) as a way to generate new financing sources had mixed success. PPP was implemented for Radarsat in Canada and Galileo in Europe, but the complexity of the deals, and the disputes about ownership, caused many problems.

In the late 1990s, private capital was available for space ventures. However, the spectacular collapse of the ambitious telecommunications constellations such as Iridium, Globalstar, ICO, Orbcomm, and Teledesic practically froze access to capital. As discussed earlier, there are few space enthusiasts who are motivated and wealthy enough to invest in private space exploration. However the lack of conventional early stage financing, such as angel investors, venture capital or private equity, makes the life of many space entrepreneurs very difficult.

Inadequate Use of Marketing Tools

Successfully promoting space activities requires the mastery of marketing and public outreach, two disciplines that are quite far from the core technical expertise readily found in the space industry. A marketing strategy is composed of four main elements, also known as the "4Ps" of marketing. These elements are Product, Price, Promotion and Physical Distribution. Managing these four elements in an integrated fashion and achieving the optimal balance between them is the essence of marketing. In other words, a successful product or service has to meet the quality and functionality expectations of the market (Product) at a fair, market-driven and competitive price (Price). Moreover, the customer must be informed about the availability and features of this offering (Promotion) and the product must be brought, via various channels, to the customer (Physical Distribution).

The nature of the product or service will largely determine the optimal marketing "mix." For example, for marketing smartphones, product characteristics such as the operating system, screen size, screen quality, connectivity and battery life are likely to be emphasized; for marketing cosmetic products, promotion is more likely to be the main focus.

Although marketing played a central role in the evolution of consumer products and services, it has been largely absent from the space industry until the 1990s. As discussed earlier, close ties to the defense industry (especially during the Cold War era) significantly hampered the efforts to openly communicate the merits of space technologies for fear of divulging critical military and commercial secrets. Increased globalization and commercialization in the 1990s eased up some of these restrictions and resulted in the introduction of systematic marketing into the space sector.

A fifth element of marketing, Philosophy, has also been proposed as part of an optimal marketing mix for promoting space activities.[1] Especially space exploration and space science missions have a natural propensity to intrigue the general public, as these missions are directly linked to many fundamental questions such as the origin of the universe, the search for other forms of life in our Solar System, risks posed by NEOs and the impact of climate change. Thus, embracing the philosophical rationale for space exploration can be particularly useful in better communicating the benefits of space activities and creating a sustainable base of public support.

Survey results indicate that the American public has historically supported the U. S. space program, even though they are not very familiar with its details. Since the beginning of NASA's space activities, various polls have shown that more than 60 % of those polled rated NASA's performance as either "excellent" or "good." However, the same respondents also indicated that federal funds could be better spent on other programs such as national defense, anti-poverty programs, education and health care.[2] A similar poll conducted in Europe in 2009 has shown that, although there is overall support for European space activities (even during a severe financial crisis), the public is divided when it comes to the level of funding. About 43 % of the respondents indicated their preference for an unchanged budget, while 23 %

preferred a budget cut. Only 20 % of the respondents opted for a budget increase.[3] As these survey results indicate, the support of the public is not always a given, and both the space agencies and the industry have to make a concerted effort to better communicate the benefits and costs of space programs.

Public Outreach

Space agencies as well as companies in the space sector have struggled with communications and public relations since the very beginning. Although some of the key events in space history, such as the *Apollo 11* and Mars *Pathfinder* landings, have generated enormous public interest, sustaining this interest and converting it into bigger investments has been a big challenge. A catastrophic event, such as the *Challenger* and *Columbia* accidents, seems to cause a sudden peak in public interest in space, albeit not in a very positive way.

Space exploration has always been an attention grabber, especially during dramatic moments. Some of the historical milestones in space history broke records in media coverage. It is estimated that 538 million people watched the grainy image of Neil Armstrong's first steps on the lunar surface in July 1969 on TV, marking the first truly global satellite broadcast and the largest television audience ever to that point in history.[4] Interestingly, this broadcast wouldn't have been possible if the Intelsat global satellite network had not been in place. By moving a satellite over the Indian Ocean just weeks before the landing, Intelsat managed to cover the entire globe with satellites and enabled the excitement of the lunar landings to be shared worldwide. Another piece of history was written in 1997 as NASA's *Pathfinder* mission touched down on Mars. Images taken on Mars broke Internet records. Various Pathfinder websites attracted more than 500 million web hits within a month.

In recent years new public outreach channels have emerged. One of these channels is participatory exploration. It enables the active involvement of citizens as contributors to space research, science, and exploration activities. Participatory exploration is not a one-way communication or outreach activity but an interactive process in which the general public can contribute by submitting ideas and creative concepts that can be incorporated into future missions.

Globalization and Consolidation

Space activities are becoming increasingly global, as new countries acquire the necessary know-how to gain access to space. Many emerging space nations have focused on microsatellites as a stepping stone for more sophisticated space activities. With the help of such entities as Surrey Space Technology Ltd. (SSTL) and Utah State University, a growing number of countries have deployed some rather sophisticated microsatellites. Learning about the microsatellite design and

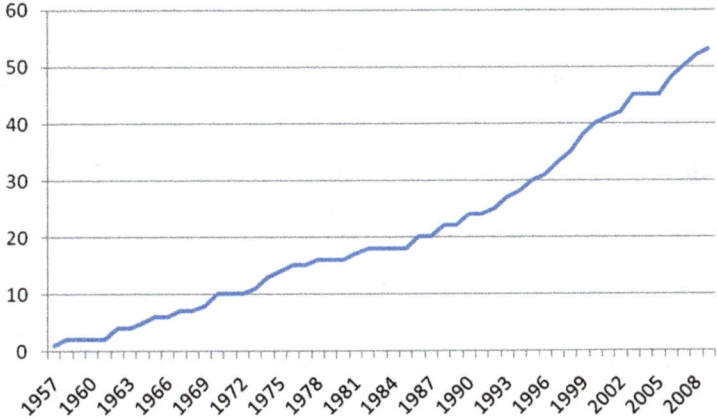

Fig. 6.1 A historical perspective – the number of countries owning a satellite (Data Source: UCS Satellite Database)

manufacturing process has enabled dozens of nations and organizations to acquire critical know-how and start pursuing more ambitious space missions. The Disaster Monitoring Constellation, a LEO constellation composed of several microsatellites (each belonging to a different country), is a testament to the capabilities of microsatellites as a force in the globalization of space activities.

Companies with increased economies of scale were able to take advantage of global business opportunities (especially in satellite manufacturing), by reaching out to customers around the world. Although there are still many export restrictions that inhibit the spread of key satellite technologies, more than 50 countries are now capable of owning and/or operating a satellite. The evolution of satellite ownership over the years gives us a glimpse of the rate of globalization in the industry. Since the beginning of the 1990s, the diffusion of satellite technologies has accelerated (Fig. 6.1).

One recurrent theme in the last few decades has been the consolidation of companies through acquisitions and mergers. Maturing technology, risk of government budget cuts and worldwide business opportunities are only some of the developments that rewarded a more global and collaborative way to do business, triggering a wave of consolidations, mergers and strategic alliances.

At the beginning of 1980s, there were about twenty major space companies in the United States. After successive rounds of mergers and acquisitions, only three major aerospace corporations were left by 1997 – namely Boeing, Lockheed Martin and Northrop Grumman. Companies in Europe also went through a similar transformation in the 1990s, leading to the emergence of just two major space conglomerates that are currently operating at the prime contractor level – namely European Aeronautic Defence and Space company (EADS)/Astrium and Thales Alenia Space.

From a public policy point of view, industry consolidation is a delicate issue. It can create many benefits, such as the emergence of global leaders with end-to-end capabilities, better access to capital markets and risk diversification across different market segments, but it can also stifle innovation in the marketplace by limiting competition and generating monopolistic behavior. Certainly this issue is not limited to the space industry, and many other capital-intensive industries also show similar evolutionary paths.

The trend toward consolidation and market-driven mergers is not necessarily a global phenomenon. In countries such as Japan, China and Russia, the composition of the space industry was steady over the past two decades. In Japan, there is a very clear division of roles between various contractors when it comes to subsystem design and manufacture. In China, virtually all space projects are carried out by the Chinese Aerospace Science and Technology Corporation (CASC), closely following the priorities set by the government. In Russia, declining space budgets after the disintegration of the Soviet Union caused a major shift in business practices and forced Russian companies to find alternative sources of revenues, including joint ventures with Western companies.

In general, globalization has dramatically altered space business, and created a need for a new generation of space professionals who are well-versed in the international dimension and cultural aspects of doing business in different countries.

The Changing Role of the Private Sector

During the early years of the Space Age, the private sector took on the role of government contractors with little or no in-house initiatives to develop new products and services. Largely responding to the needs of governments in the civilian and military domains (and fueled by a race between the United States and the former USSR), companies were not very proactive in exploring new markets.

Since the beginning of 1990s, however, a major trend in space activities has been decreasing or leveling public funding (with the possible exception of China and a few other countries). In contrast, the private sector increased its investments in space. Although there were some disruptions along the way, such as the bursting of the dot.com bubble and the attacks of September 11, 2001, this trend is still largely intact. In fact, the Great Recession in the United States and Europe (from 2008 through 2012) has resulted in the transfer of debt from the private sector to the public sector, and a new age of austerity is on the horizon on both sides of the Atlantic. The reduced spending power of the governments will no doubt put additional strain on the already limited budgets of the space agencies (as evidenced by NASA's recent decision to pull out from joint Mars missions with ESA).

This is not necessarily a bad development. Budget constraints may force space agencies to focus on what they do best – leading technology development in cutting-edge exploration projects and satellite applications, while the private sector can focus on perfecting the infrastructure supporting these activities (including access to space). This can also allow developing innovative commercial applications to take full advantage of both public and private investment. The U. S. policy enacted through the Commercial Space Transportation Act, the Commercial Space

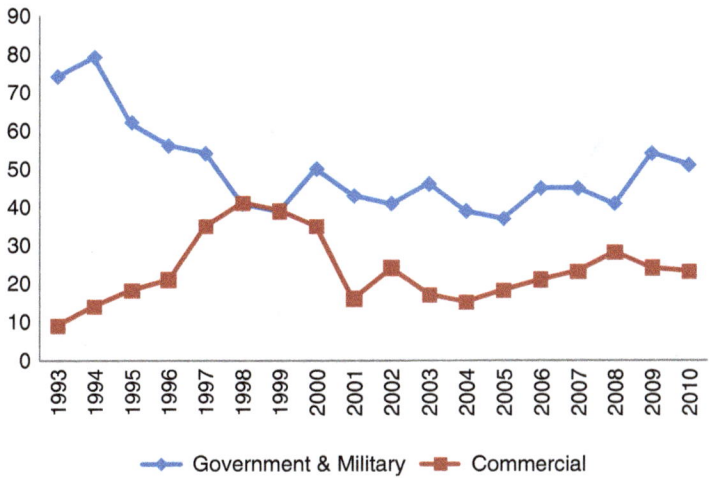

Fig. 6.2 Public and private sector launch events around the world between 1993 and 2010 (Data Source: FAA)

Amendment Act of 2004 and the last amendment passed at the end of 2011 and signed into law in January 2012 are all illustrative cases in point.

An interesting indicator that highlights the changing roles in the space industry is the ratio of commercial launches to military/government launches. Although this is a somewhat limited metric, as it doesn't capture the economic value of each launch, it can still be used as an objective indicator that is consistent over time.[5] As can be seen in Fig. 6.2, the number of commercial launches worldwide peaked in 1998 with 41 launches. Signifying the height of the dot.com boom, the demand for these launches were largely from the LEO telecommunications constellations. The entire Iridium constellation of 66 satellites plus spares was launched within a record-breaking 13-month period.[6] In one 13-day window (from late-March to early-April 1998) 14 Iridium satellites were successfully placed into orbit.

In stark contrast to the boom of the late 1990s, the turn of the century was marked by a collapse. During 2001, only 16 commercial launches took place. Since then, the number of commercial launches has increased modestly, but it is still significantly below the volume achieved in the late 1990s.

Another interesting trend that can be deduced from Fig. 6.2 is the decline of military launches throughout the 1990s. The end of the Cold War freed up ample launch capacity (especially in Russia) that has been re-channeled to commercial launches.

While the commercial space market was going through its cycles, military space expenditures also have had their ups and downs. At the end of the Cold War, DOD's space budget peaked at around $29 billion. Throughout the 1990s, as the defense budgets contracted across the board, military space activities in the United States were not spared. However, especially in the United States, military space expenditures increased significantly during the last decade following the attacks of September 11th

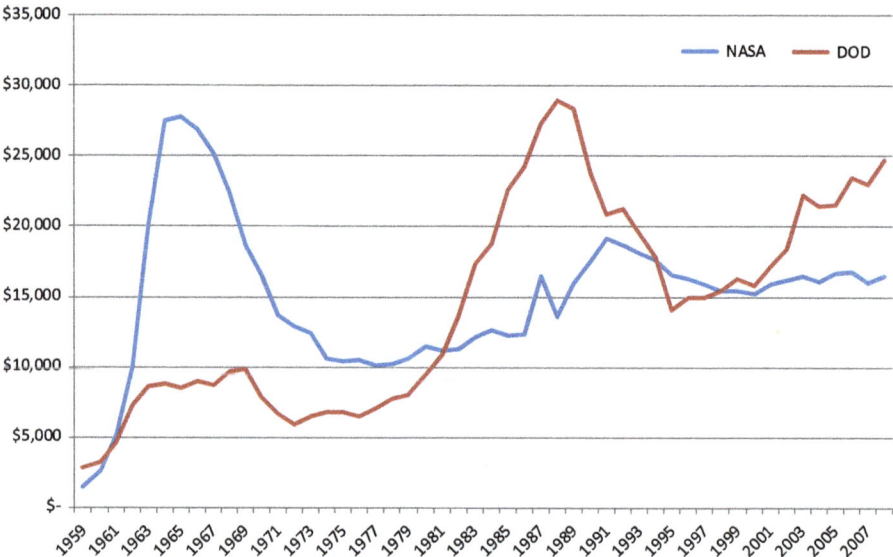

Fig. 6.3 Evolution of NASA's and DOD's space budgets over time (in $ millions adjusted for 2008) (Data Source: NASA Aeronautics and Space Report of the President)

and the wars in Iraq and Afghanistan. Much of this increased military expenditure has been directed towards collection of intelligence and surveillance. The fact that the Department of Defense can "dual use" commercial facilities to relay data for a variety of purposes complicates the financial picture even further (Fig. 6.3).

Thus, for the last 10 years, given the relatively stable budgets of the civilian space agencies in developed economies, the overall increase in the government expenditures were largely due to the military programs. This recent wave of investment in military space technologies may eventually result in many new civilian applications and generate new business opportunities.

Space Debris

Space debris can be seen as the ultimate negative externality caused by space activities. After decades of launches and spacecraft operations, Earth's orbit is full of space "junk" that can cause very serious damage to space missions. An avalanche is a good analogy to illustrate the magnitude of this problem. After a tipping point, the accumulated debris may start colliding with each other, creating even more debris. This cascading effect can take out many operational satellites, and perhaps more importantly, it can render certain parts of the orbit unusable for many years until orbital perturbations naturally clean some parts of the orbit or engineering solutions are developed to clean up the debris.

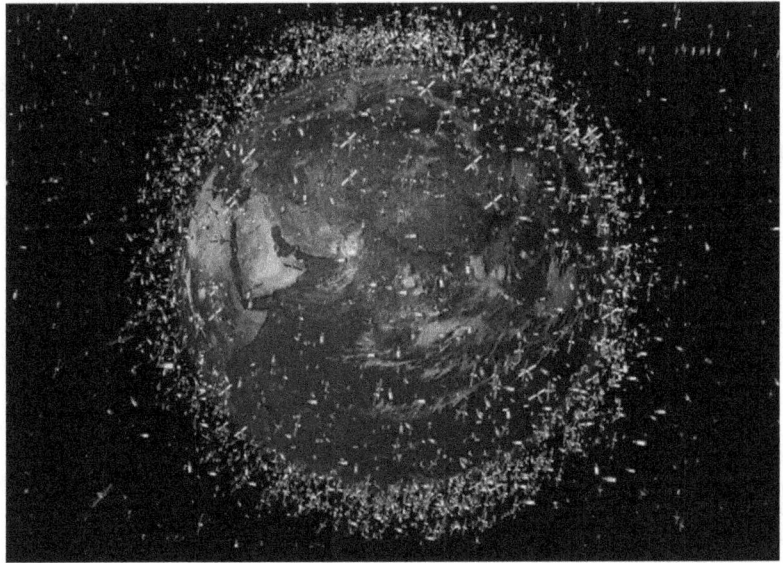

Fig. 6.4 The space debris field around LEO (artist's impression) (Source: ESA)

Some of the developments discussed earlier, such as the proliferation of microsatellites, cubesats and other smaller spacecraft in orbit also exacerbates the space debris problem.

Based on the risk mitigation strategies explained in Chap. 7, there are a few ways to prepare for a space debris event, including buying insurance, adding protective armor to satellites to withstand debris under 1 cm in size, or space situational awareness so that active maneuvers can be used to avoid anticipated collisions. In the future, however, active debris removal could represent a new business opportunity. This subject is addressed in detail in the book "Space Threats" that is a part of the Springer Briefs in Space Development series on behalf the International Space University (Fig. 6.4).

Replacing Generation Apollo

The launch of Sputnik in 1975 was a wake-up call for the young generation in the United States. The interest in science, engineering and mathematics education spiked, and within a single decade a highly motivated wave of employees entered the space workforce. The average age of NASA employees during the Apollo program was about 30 years. Currently, the average age of NASA personnel is 46, and 68 % of its population is between 35 and 55.[7]

"Generation Apollo" powered NASA's various space programs either as NASA employees or contractors. Today, most of these workers are reaching retirement age; therefore, preserving their know-how and building the next generation workforce are important priorities. Although the situation is not as dire in Europe, Canada and Japan, all spacefaring nations need to make sure that important lessons and best practices are not lost between generations. To this end, NASA, ESA and JAXA have invested in knowledge management tools to make sure that critical organizational knowledge is not lost.[8]

Notes

1. Peeters, W.A.R, Space Marketing: A European Perspective, Kluwer Academic Publishers, Dordrecht, The Netherlands, 2000.
2. Launius, R. D., "Public Opinion Polls And Perceptions Of US Human Spaceflight", Space Policy, vol.19, pp. 163–175, 2003.
3. Eurobarometer, Space Activities of the European Union, Conducted by The Gallup Organization, Hungry, 2009, accessible at http://ec.europa.eu/public_opinion/flash/fl_272_sum_en.pdf.
4. Pelton, J. and Bukley, A. (eds.), "The Farthest Shore: A 21st Century Guide to Space", Apogee Books Space Series, 2010.
5. We should note another limitation of the number of launches as a metric of economic activity: the current generation of launchers can carry heavier payloads (up to around 22,000 lb, or 10,000 kg to LEO) and are lifting two or more satellites into orbit at the same time. Furthermore, today's satellites are also more capable: a few decades ago, a typical communications satellite had 12–24 transponders on board, Today the capacity of a telecommunications or broadcasting satellite might be four to six times larger (for example, some of the Intelsat satellites have had as many as 100 transponders on board). In short fewer launchers does not always mean that less payload goes into orbit.
6. Another impressive fact about Iridium: the satellites were launched by rockets from the US, Russia and China: a wonderful example of the international nature of space business.
7. Topousis, Daria E., "Multi-Generational Knowledge Sharing For Nasa Engineers", Proceedings of IAC 2009, Korea, 2009.
8. For more information about these efforts, please see "Knowledge Capitalization in a Concurrent Engineering Environment" by Schubert, D. et al, proceedings of the 61st International Astronautical Congress 2010, Prague, Czech Republic.

Chapter 7
Risk Management

One afternoon in 1952, at the library of University of Chicago, a young researcher had a Eureka moment. As part of his PhD dissertation, Harry Markowitz was trying to find a way to construct optimal portfolios of financial securities for managing investment risk. Since individuals have different levels of tolerance for risk, Markowitz was looking for a way to build portfolios such that for a given level of risk tolerance, the expected return is maximized.[1] If an individual has a higher risk tolerance, then he or she should be rewarded by a potentially higher return.

For Markowitz, the particular challenge was to define and measure investment risk. His initial insight was to use the variance (dispersion around the mean) of financial securities as a measure of risk. Although this seemed like a fairly reliable metric to measure the risk of an individual investment, simply adding the variance of each security would not accurately represent a portfolio's total risk. His second insight was to find a way to map the relationship between different types of investments to compute the risk of an entire portfolio. For this purpose, he computed the statistical relationship between each security (the covariance) and used this relationship to achieve risk diversification.

By computing the risk of an entire portfolio, and by taking into account both the overall investment risk and return simultaneously, Markowitz demonstrated that optimal risk-return combinations can be created. As we will see shortly, the applications of his work have spilled over from finance into many other fields including medicine, defense and space. Decades after the seminal work of Markowitz, risk management has now become a discipline of its own. In 1990, Markowitz was awarded the Nobel Prize in Economics.

Defining Risk

At the most basic level, risk refers to both the likelihood and the negative consequences of an event. Although this definition is very straightforward, in practice, most of us fail to take into account both of these factors in a rational way before making our decisions. Even if we may have complete information regarding the likelihood and consequences of an event, various factors such as emotions and time pressure can wreak havoc on the accuracy of our decisions.

For example, our fear of catastrophes such as plane crashes, suicide bombers and shark attacks cause us to overestimate the odds of these terrible but infrequent events. On the other hand, we routinely underestimate the risk of ordinary events, such as driving on the highway. This warped perception of risk can result in unexpected consequences. After the attacks on September 11th, many people opted for driving instead of taking commercial flights. However, it is a well established statistical fact that driving is much more risky than taking a commercial flight. Past data show that driving the length of a typical non-stop flight is about 65 times as risky as flying on a major U. S. airline.[2] Not surprisingly, as more people decided to hit the road, the number of fatalities increased. Researchers estimate that this change in travel preferences resulted in more than a thousand additional fatalities since September 11th.[3]

Risk in Space

Given the harsh space environment, it is only natural that space is a risky business. Many planetary missions, including Mars Polar Lander, Phobos Grunt and Beagle-2, failed at various stages after their launch. Although these losses have resulted in lost science opportunities, the impact of a catastrophic accident during a human spaceflight mission goes much deeper. From the astronauts of *Apollo 1, Challenger* (STS-51), *Columbia* (STS-107) to the cosmonauts of *Soyuz 1* and *Soyuz 11,* many explorers made the ultimate sacrifice to advance the cause of space exploration.

Fortunately, compared to the historical track record of mission failures, great advances have been made in containing and managing risk. At the beginning of the Space Age, before the introduction of systems engineering, along with other related innovations such as redundancy and environmental testing, failure rates were around 50 %. Today, failure rates are around 5–10 % for robotic missions and even lower for human spaceflight (Fig. 7.1).

NASA's more formal definition of risk is "a measure of the inability to achieve overall program objectives within defined cost, schedule, and technical constraints."[3] Risk in space projects involves two main components: the probability of failing to achieve a particular outcome and the consequences/impacts of failing to achieve that outcome.

This rather sterile definition equally applies to a catastrophic outcome (e.g., a fatal spacecraft accident) as well as a minor one (e.g., a 5 % cost overrun). The key is to "map" the likelihood against the consequence. For example, an event with a relatively low risk can rank very highly in a "risk list," if the consequences are catastrophic. Commercial airline accidents are in this category, as well as the threat of a large-sized NEO colliding with Earth.

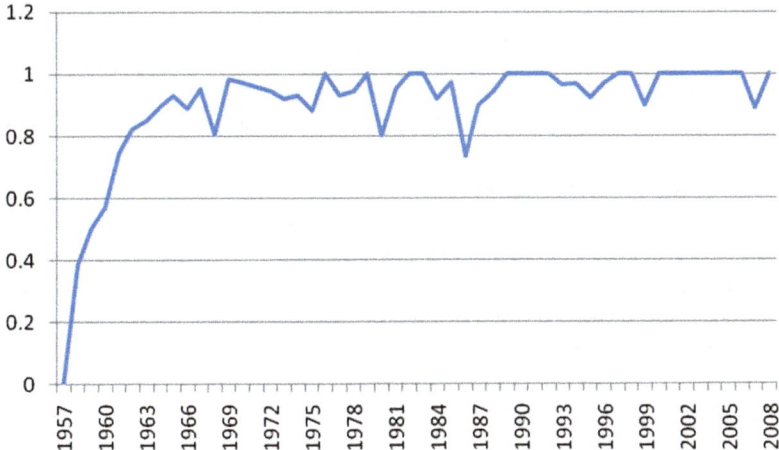

Fig. 7.1 Ratio of successful missions to total number of missions since 1957. (Data Source: NASA)

Types of Risks

Risks in space projects stem from a variety of sources. The main categories are described below.

- Cost Risk: As discussed in the next chapter, space programs have a poor record when it comes to the accuracy of cost estimates. Cost risk encompasses two main potential threats: inaccurate cost estimates and cost overruns due to schedule slippages, technical problems or requirement changes.
- Schedule Risk: Time is money in space project management. Just like cost risk, inaccurate estimates of a project's duration or schedule slippages can pose a significant challenge and threaten the performance of space projects. In many cases the result is a cost overrun. In some cases, meeting a specific launch date can be crucial, such as an Earth-Mars launch window.
- Technical Risk: From the general public's point of view, this is probably the most evident type of risk. History of space exploration is full of catastrophic accidents, explosions at the launch pad and lost missions due to technical malfunction of hardware or software.
- Programmatic Risk: Sometimes policy-related and legal developments beyond the control of the project manager can adversely affect a space mission. Typical examples include the risks posed by regulations (e.g., the International Traffic in Arms Requirements) and changes in policy imperatives.
- Market Risk: Although it may not be a primary concern for space agencies, market risk is a significant issue for the private sector. Even though a space project may work perfectly within the projected budget and schedule, any significant

changes in the demand side can have very adverse consequences for business. The Iridium case is an excellent, albeit painful, reminder of market risk.

It should also be noted that human error can also cause devastating results. A famous case is that of the Mars Climate Orbiter, a spacecraft that disintegrated during orbit insertion around Mars in 1999. A subsequent investigation revealed the root cause of the problem: the on-board software of the orbiter was designed to process thrust instructions using the metric unit Newtons (N), while ground control generated those instructions using the Imperial measure pound-force (lbf).[4] Thus, although the spacecraft perfectly executed all the instructions without a technical malfunction, a slight omission in converting physical units on a single line of software code inadvertently caused a total loss of mission.

Modeling Risk

Keeping track of all these types of risks in a space project is a challenging endeavor. Mapping the likelihood and consequences of various risks is an arduous task by itself. On top of that, we also need to model the impact of one risk on another. For example, a technical risk, such as the solar panels of a telecommunications satellite not deploying during commissioning, can result in severe market risk, such as loss of revenue.

In order to quantify and analyze various risk factors, we have various tools at our disposal. Some of the most basic, and useful, tools are the risk matrix and event trees.

Risk Matrix

A risk matrix is a table that can be used to analyze the likelihood and consequence of multiple risk factors. It can be very useful in prioritizing the risks of a project. This prioritization enables the project manager to make efficient use of the scarce resources available for managing risk. It can also be a powerful tool for communicating these risks to various project stakeholders (Fig. 7.2).

If not properly managed, the risk factors that are categorized as high priority can have devastating consequences. Therefore they almost always require additional actions. Risks in the moderate category need to be monitored closely. Limited resources may require a project manager to accept these risks and take no further action. The risks in the low priority category may look docile, but this can be misleading. The interaction of these risk factors with other ones may significantly amplify their consequence. In fact, this is one of the biggest limitations of a risk matrix. The insight of Markowitz, as discussed in the introductory part of this chapter, is applicable in this case. Any holistic approach to risk management requires an analysis of the interactions of the risk factors.

Modeling Risk

Fig. 7.2 A risk matrix (Source: NASA)

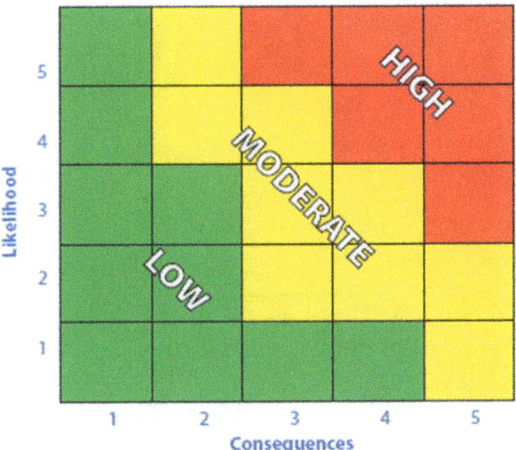

A heartbreaking reminder of these interactions took place in 2003. Space shuttle *Columbia* disintegrated during re-entry, taking the lives of the seven astronauts onboard. This accident was caused by an incident regularly observed in previous shuttle flights, namely the shedding of a foam particle from the external tank insulation during lift-off. The accident investigation revealed that the NASA management had known about this issue for many years. However, since it didn't cause a devastating effect in previous shuttle flights, it was categorized as a low/moderate priority risk, and no comprehensive action was taken. Unfortunately, by damaging the leading edge of one of *Columbia*'s wings, a piece of foam cracked the thermal shield of the spacecraft. Without sufficient thermal protection, the crew had no chance for survival during the re-entry phase.

Event Tree

An event tree is a graphical representation of how risks and their possible consequences are linked, including the outcomes of chance events or states of nature. A variant of the event tree is a decision tree, in which various design and development decisions can also be included in the analysis.

Figure 7.3 illustrates how an event tree can be used to analyze these interactions. In this example, we assume that a Mars sample return mission has a 75 % chance of landing successfully on the Red Planet. After landing, the spacecraft is tasked with collecting samples and bringing them back to Earth, with a 33 % chance of success. The event tree shows all the three possible outcomes of this mission: successful landing and return, successful landing and failed return, and finally, failed landing. Moreover, it also shows that the probability of overall mission success in this hypothetical example is 25 %.

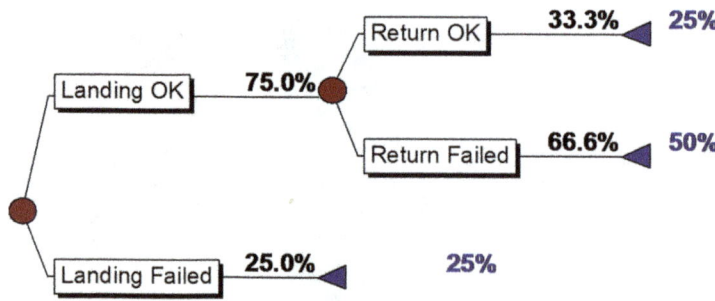

Fig. 7.3 An event tree

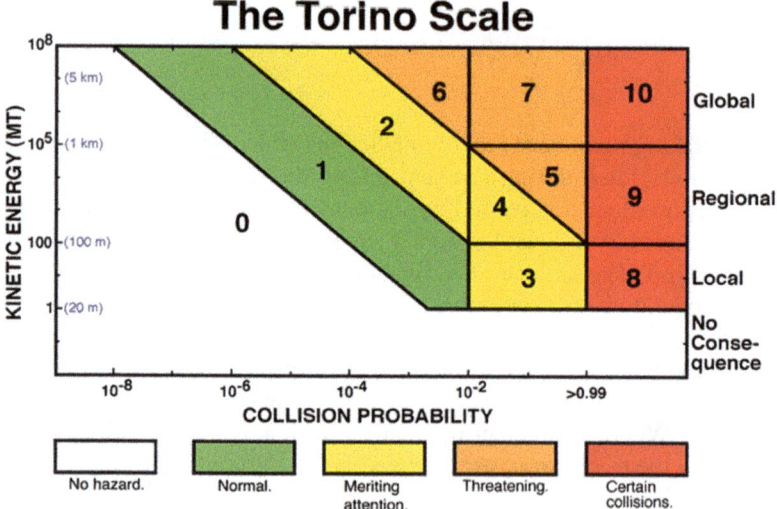

Fig. 7.4 The Torino Scale (Source: NASA)

Hazard Scales

A hazard scale is primarily used as a way to communicate the risk of a particular event to the general public. Similar to the risk matrix, a hazard scale is also based on the combined impact of the likelihood and consequence of various risk factors. With the additional step of ranking the outcomes, a scale is generated. For example, NASA started to use the Torino Impact Hazard Scale in order to better communicate the risk profile of various NEOs.[5] Developed by a researcher from MIT, this scale takes into account the object's size and speed as well as the probability of impact with Earth. In this example, the size and speed of the object determines the consequence of the risky event, while the probability of impact is used as the likelihood measure (Fig. 7.4).

The top right portion of the scale denotes high risk areas. A scale value of 5 or more would certainly not be good news, as it would mean a close encounter with an object that is large enough to cause regional or global devastation.

Conversely, an event that may not have a devastating consequence but a very high chance of occurring repeatedly may also cause concern. The cumulative impact of these low impact events may add up and start posing a significant threat (the degradation of solar panels in space, or the cumulative effect of colliding with low-mass space debris). Between the two ends of this spectrum there are many other likelihood and consequence combinations that can threaten space activities.

Managing Risk

Identifying and mapping risks may be very useful in understanding the challenges a project faces, but the real work starts after this stage. We now need to manage the risks we have identified. Risk management is one of the key skills required to survive in the challenging environment of space. However, it should not be construed as the total elimination of risk, as this is simply not possible. After all, taking calculated risks is essential for progress in both exploration and business.

Generally speaking, there are two main avenues we can take. We can actively try reducing the risk by mitigating it, or we can purchase insurance to offset the financial losses. It is very important to realize that these two avenues are not mutually exclusive; we can invest in a certain amount of redundancy to mitigate the risk, and then transfer the rest of the risk through an insurance product.

Risk mitigation is comprised of engineering and management practices that can be implemented to decrease the likelihood and/or negative impacts of various risk factors. Some of the specific tools in our arsenal include testing, redundancy and mission-level diversification.

Testing and Redundancy

Many seasoned aerospace engineers would give the same advice to young engineers: "Test early. Test often. And then test again." Complex missions can have thousands of failure modes, and extensive testing and simulation can reveal many hazards along the way. Once in orbit, most satellites are beyond the physical reach of their owners to conduct and repair or upgrade work. Thus, building in redundancy is another key tool to minimize risk. If a back-up is in place, mission controllers have a much higher chance to manage in-orbit failures.

Diversification: Towards a Portfolio-Based Approach

An organization's attitude towards risk largely determines the nature of space missions that get funded. For example, in the early 1990s, NASA administrator Dan Goldin introduced a strategy dubbed "Faster, Better, Cheaper," encouraging engineers and designers to take more risks for a given mission, but diversifying the risks across multiple missions. Such an approach is particularly suitable for robotic exploration missions, for which the consequence of losing a single mission is not catastrophic. Thus, low-risk and high-risk missions can be combined in a portfolio of missions, akin to the financial portfolios discussed at the beginning of the chapter. For example, the Spirit and Opportunity rovers were not necessarily cheap, but they demonstrated the phenomenal upside in space missions when excellent engineering and program management skills are paired up with a healthy dose of luck. Building the same rover twice may not look as cutting-edge as some may desire, but it sure provides excellent science.

On the other hand, it is significantly more difficult to manage human spaceflight missions as portfolios. The consequence of losing the crew during a mission is catastrophic, not just for the families of the astronauts but also for the image of the space agency involved.

Risk Transfer and Insurance

Given that it is not possible to reduce mission risks to zero, most commercial satellite operators routinely use insurance products as part of their risk management plans. In case of a launch or in-orbit failure, insurance can be a very useful risk management tool to limit the financial losses. However, making a claim and receiving financial compensation from the insurance company doesn't bring back the lost spacecraft. Service interruption, lost revenues and damaged company reputation are also distinct risks that have to be managed.

Although it is quite common to use insurance in the commercial segments of the industry most other segments rely on the pockets of the biggest customer, the government, as their last resort. When governments guarantee a space mission, there could be a particularly strategic or scientific rationale. Some of the largest service providers such as Intelsat have become self insurers, but such behavior is rare and only occurs when insurance premiums seem to be at especially elevated levels.

There can also be very creative ways to use insurance. For example, the ten million dollar Ansari X-Prize was kick-started with an insurance product. The financial commitment of the Ansari family was limited to the insurance premium, not the prize itself, which was paid by the insurance company.

Case Study: Nuts and Bolts of Risk Management

Failures in space activities do not always happen in space. In 2003, a $240-million satellite, NOAA N-Prime, toppled over and crashed to the floor of the facility where it was being assembled. The crash caused extensive damage to the satellite, but the

Case Study: Nuts and Bolts of Risk Management

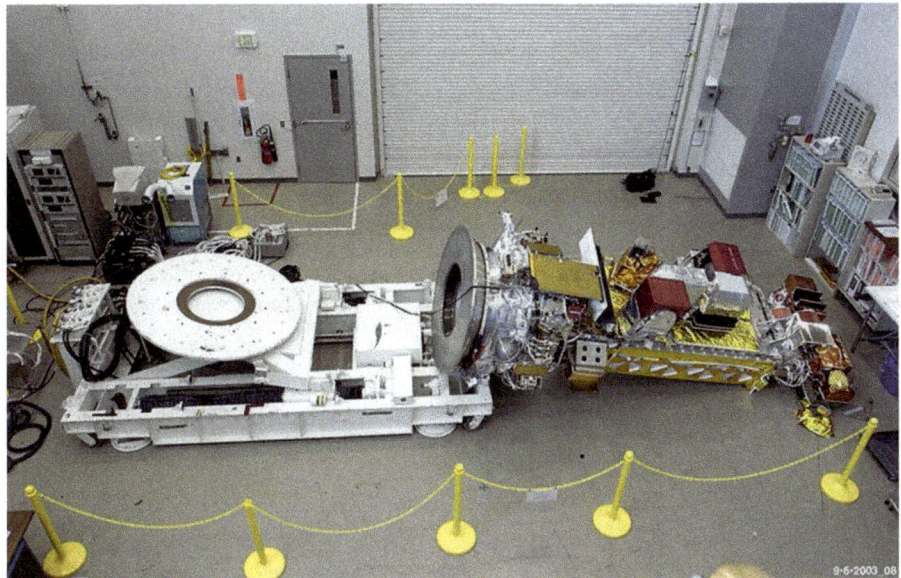

Fig. 7.5 An Ooops! moment: NOAA N-Prime satellite crashed on the floor (Source: NASA)

biggest surprise was the cause of the crash. NASA's investigation concluded that the accident was a result of 24 missing bolts that were supposed to secure the satellite as it was being rotated. It turns out the bolts were removed by a technician working on a different project, but this little detail was not communicated within the facility. Eventually, the satellite was repaired at a cost of $135 million, and it was successfully launched in 2009 (Fig. 7.5).

The seemingly trivial root cause of the NOAA N-Prime accident is not unique. Various accident investigations revealed a similar pattern. An O-ring that had shrunk in size due to the surprisingly low temperatures at the Kennedy Space Center at the time of launch caused the *Challenger* accident; loose insulation foams were to blame for the *Columbia* accident; and a metric to imperial unit conversion error claimed Mars Climate Orbiter. However, many experts believe that these technical malfunctions are symptomatic of a deeper ailment related to the prevailing culture of an organization. The mission control team decided to launch *Challenger* on a morning with freezing temperatures, NASA management knew about the insulation foam problem but categorized it as an acceptable risk, and finally, a breakdown in communication between different team members and the inability of management to connect the dots caused the metric-to-imperial conversion error go unnoticed.

Notes

1. An equally valid formulation could be to minimize the level of risk for a given rate of return.
2. Sivak, M. and Flannagan, M., "Flying and Driving After The September 11 Attacks", University Of Michigan Transportation Research Institute Research Review, July–September 2002, Volume 33, Number 3, available at: http://www.umtri.umich.edu/content/rr33_3.pdf.
3. Blalock, G., Kadiyali, V., and Simon, D. H., "Driving Fatalities After 9/11: a Hidden Cost Of Terrorism", *Applied Economics*, Volume: 41, Issue: 14, 2009.
4. NASA, "Mars Climate Orbiter Mishap Investigation Board Phase I Report", November 1999, available at ftp://ftp.hq.nasa.gov/pub/pao/reports/1999/MCO_report.pdf.
5. For more information on this scale, see http://neo.jpl.nasa.gov/torino_scale.html.

Chapter 8
Cost Management

In November 2005, at a U. N.-sponsored conference, Nicholas Negroponte, co-founder of the MIT Media Lab, unveiled the prototype of a $100 laptop computer. Dubbed "The Children's Machine," this computer was designed to be a low-cost alternative to mainstream computers, making it accessible to the children of the developing world. At the time of this announcement, personal computer prices were around $1,000, still beyond the purchasing power of families from many developing countries. Since then, a remarkable thing happened: computer prices decreased significantly, and a new generation of computing devices (netbooks, tablets and smartphones) burst into the global market. Negroponte's dream has been realized within less than a decade and millions of laptops reached the hands of schoolchildren around the world (at a price of about $200 per unit). Although large orders given by governments were crucial for success, the delicate dance between the demand and supply in the IT sector was the main driver behind the mass production of low-cost laptops.

In October 1957, the world's first artificial satellite, Sputnik, was launched aboard an R-7 rocket. Derived from an intercontinental ballistic missile, this rocket evolved into the workhorse of the Soviet space program, the Soyuz launch vehicle. Although we cannot easily estimate the cost of the Soyuz launches in the earlier days (due to the secrecy of the Soviet space program), we can safely say that the cost has not come down significantly despite half a century of experience in designing, building and operating similar rockets. The recent foray of SpaceX into the launch vehicle business may be the tipping point towards lower prices, but more launch campaigns are needed to verify the reliability and profitability of this new venture.

Many publicly funded projects face cost overruns, but some of the examples in the space industry are staggering. The original concept for a space station originated during the Reagan presidency (named "Space Station Freedom") envisioned that the project would be completed by 1992 and cost about $8 billion. The ISS, successor of Space Station Freedom, was completed in 2011. The life-cycle cost of the ISS is estimated at $175 billion, including the cost of all the shuttle missions required to assemble it.

So, how can we explain drastic cost reductions in the IT sector within a single decade versus the stubborn cost of space missions?

Cost Analysis and Management

Cost analysis and management is one of the core topics of systems engineering. Almost all space projects are designed, built, and operated by striking the best balance possible between performance, cost, schedule, and risk. These "levers" of space mission design can be explained very simply: "You can have it cheap, you can have it quick and you can have it best. But you cannot have it all at the same time."[1]

During the space race, cost and risk reduction took a back seat, as the United States and the USSR poured resources into their space programs in order to maximize performance and minimize the schedule. The reality of the space industry is very different now, and managing the cost of space projects in the most efficient way has become a priority.

Complexity of space projects and the high cost of access to space are the main reasons for the industry's high cost. The space industry is by no means the only sector of economic activity with this particular problem; nuclear power plants, military projects and large-scale infrastructure investments are also prone to significant cost overruns. History is full of megaprojects with ballooning price tags: the Suez Canal cost nearly 20 times more than the original estimate, the Sydney Opera House nearly 15 times and the Concorde passenger jet 12 times.[2]

A key concept in cost analysis and management is the life-cycle cost, defined by NASA as the total cost of ownership over the system's life cycle, including feasibility, design, manufacturing, deployment, operations and disposal efforts.[5] In many cases, space agencies report "baseline" budgets which do not include the cost of operating and disposing of the systems. However, operations is one of the most expensive phases of a space mission, as it involves the ongoing labor expenses of highly qualified personnel as well as the use of ground facilities. A historical analysis of NASA's missions shows that life-cycle costs can be dramatically more than the baseline costs. Therefore, when any type of cost analysis is performed, it is critical to distinguish between the baseline figures and the full life-cycle estimates.

Another key issue is the distinction between recurring and non-recurring costs. Initial R&D efforts and ground infrastructure are non-recurring costs that can be spread over many missions. Recurring costs, on the other hand, are generally mission specific and incurred each time a mission is launched and operated. In most industries, non-recurring costs are spread over millions of production units, and thus the cost per unit of a product is minimized. As discussed earlier, space manufacturing is a very low volume practice resulting in eye-popping unit costs.

The cost per launch of the space shuttle is an illustrative case. Calculating the exact cost has been elusive until the very end of the program. Billed as a reusable vehicle that can fly multiple missions with reasonable maintenance costs, the shuttle never achieved the cost efficiency envisioned at the beginning of its life. During the design and development phase between 1971 and 1980, NASA spent about $33.4 billion (in 2010 dollars) in the form of non-recurring costs. With the initial flight in 1981, operations began. During the 31 years of operational service, 134 shuttle missions were flown at a total cost of $159.6 billion (again in 2010 dollars).[3] Thus, the cost per flight of the shuttle came to $1.44 billion, a far cry from NASA's reported $450 million. Interestingly, the actual cost of building additional units was small compared to the operational costs. The replacement of the *Challenger*, *Endeavour*, cost about $3.5 billion (in 2010 dollars). The operating and maintenance costs of the fleet, on the other hand, were staggering. The annual operating costs have been estimated at $3.8–4 billion (see Chap. 2, endnote 1).

A History of Cost Overruns

In 2004, the U. S. Congressional Budget Office (CBO) performed a detailed analysis of NASA's historical record in cost estimates. After analyzing 72 NASA missions that took place between 1977 and 2004, it was found that the average cost overrun was about 45 % (excluding the effects of inflation). Only 14 missions were completed within the initial budget allocation. Some of the flagship programs resulted in very significant cost overruns. The Hubble Space Telescope and the Galileo interplanetary probe ended up costing more than three times the original estimates. A separate study performed by the U. S. General Accounting Office (GAO) revealed similar findings. Of the 27 NASA missions analyzed, more than half had shown cost increases, in some cases doubling the financing requirement.[4]

Some of the distinguishing features of the space industry discussed in Chap. 3 contribute greatly to the occurrence of cost overruns. Space projects are long duration endeavors full of technical challenges. They are also prone to many design and requirement changes as the policy objectives evolve. For example, at the time of the original budget request in 1987, the ISS was envisioned as a primarily U. S.-led effort with a few international partners. The end of the Cold War created a policy imperative to include Russia into the program, significantly changing the technical and programmatic requirements, and consequently, increasing the cost. Another main reason for the cost overruns is the way estimates are made in the first place. One of the key recommendations of the GAO was to base the cost estimates on full life-cycle costs, including operations, maintenance and disposal.

Cost Estimation Methods

Clearly, estimating the exact cost of a space mission at the beginning of the development effort is a very challenging task. The good news is that, after decades of trial and error, experts have developed three main methods to tackle this task: costing by analogy, parametric costing and bottom-up costing.[5] These three methods can be used in tandem, and we can move from one to the other as the project evolves.

Costing by Analogy

Costing by analogy starts with identifying past missions with similar scope and complexity to the project at hand. The cost data from a past project can then be used to build a cost estimate to the present project.[6] It is important to note that this is a rather subjective method. Depending on the complexity of the project, experts may significantly increase or decrease the cost estimates based on the intensity of the required R&D effort. Additional adjustment factors may be needed to account for the maturity of the technology and inflation. This method is more suitable to projects with a repetitive character with plenty of historical examples.

For example, in satellite telecommunications, where there are hundreds of historical projects to draw from, this can be a very useful method. In fact, since the satellite communications contracts are awarded on a fixed-cost basis, such estimates based on analogy are crucial to assess the reasonability of a bidder's quote. The CBO used costing by analogy in 2005, as part of their cost estimates for NASA's future lunar missions. Analysis of the Apollo program's cost data resulted in estimates of around $100 billion for a single mission involving astronauts that would take place by 2020. A similar analysis was performed for robotic missions to the Moon using past data from the Viking mission and Mars Exploration rovers.[7]

For many other projects, however, costing by analogy may provide misleading results. For instance, in order to estimate the cost of a human exploration mission to Mars, we can use the Apollo program cost data as a baseline. Although this would be better than starting from scratch, the difference in performance requirements, life support systems, and mission duration would change the results very significantly.

Bottom-up Costing

Bottom-up costing is also referred to as "grass roots" or "engineering build up" costing, and it is very commonly used in the construction industry. This method is based on going through the specific tasks of a project and adding up all the cost elements that are attributable to each task. By incorporating additional expenses such as materials and overhead, we can arrive at a fairly accurate estimate of the cost. Naturally, as the project requirements and corresponding work packages change, the cost estimates have to be adjusted accordingly. Thus a high degree of precision with regard to the final design is needed to increase the accuracy of the cost estimates. Although this requirement may be met in most terrestrial projects, for most space contracts, it may simply not be possible in the early phases of project development. Especially for space missions with a high degree of technical complexity, the client and the contractor work hand in hand to complete the feasibility and design phases, and only after this stage is the design frozen.

Parametric Costing

Parametric costing relies on an extensive mathematical analysis of historical cost data with the aim of identifying the cost drivers of a project at a fairly detailed level (but not as detailed as the bottom-up costing approach). The link between the cost of a system and its variables are established using "cost estimating relationships." For example, the power requirement of a sub-system can be used as a fairly accurate indicator of its cost. As the power requirement increases, so would the cost. Other indicators include mass, volume and various performance metrics.[8] One of the main advantages of this method is the ability to perform "what if" analyses fairly quickly. Once the CERs are established and a mathematical model of the system is built, the cost impact of various changes in the design can be readily observed.

Cost of Major Space Programs

One of the frequently cited criticisms against space missions is their high cost. After all, there is an opportunity cost in investing public funds in space. Healthcare, education, infrastructure and a host of other government programs always compete with space for public funds. Thus, keeping track of costs is not only important for accountability but also for comparing the cost of space investments to that of other government investments.

Estimating the life-cycle cost of space programs can be tricky, as there is a need to compile cost information from the beginning to the end of a program, which can span decades. More importantly, budget amounts from different years have to be adjusted to the same base year in order to make an "apples to apples" comparison. For example, simply adding the annual budget of the Apollo program from 1959 to 1973 gives a total cost of $20.4 billion. However, as discussed earlier, due to time value of money, this amount would correspond to a much higher figure in today's dollars. Taking 2010 as the base year and adjusting for time value of money, the cost of Apollo suddenly becomes $109 billion (in 2010 dollars). The life-cycle cost of various space programs are shown in Table 8.1.

As shown in Table 8.2, comparing the cost of space programs to the cost of megaprojects on Earth can yield interesting results.

Contract Management: The Heritage from the Defense Industry

Given that most of the procurement in the space industry is undertaken by the government, the types of contracts that govern space activities play a key role in shaping the industry. Generally speaking, there are two main types of contracts: cost-plus

Table 8.1 Life cycle costs for space programs[a]

Space program	Period	Cost in current year ($ billion)	Inflation adjusted cost in 2010 dollars ($ billion)	Notes
Mercury	1959–1963	$0.27	$1.6	
Gemini	1962–1967	$1.28	$7.2	
Apollo	1959–1973	$20.44	$109	
Skylab	1966–1974	$2.26	$10	
Space Shuttle	1972–2012	$123.03	$193	
International Space Station	1987–2015	$58.70	$72	Not including the cost of the space shuttle missions and the contributions of the international partners (estimated at $24 billion in 2010 dollars)
Viking	1975–1983	$0.94	$4.16	
Mars Pathfinder	1996–1998	$0.27	$0.35	
Mars Exploration Rovers (spirit and opportunity)	2000-present	$0.92	$1.08	The cost of mission extensions is included in the estimates
Hubble Space Telescope	1977-present	$1.55	$5.57	If the five shuttle servicing missions are included, the cost surpasses $12 billion (in 2010 dollars)
Galileo Satellite Navigation System[b]	1999-present	$10.12	$29.05	Since the system is not operational yet, life-cycle cost estimates are uncertain. The cited cost in current year dollars is the initial estimate provided in 2000

[a]The following sources were used for compiling the table: United States General Accounting Office, Congressional Budget Office, Lafleur, C., "Cost of US Piloted Programs", The Space Review, March 2010, available at http://www.thespacereview.com/article/1579/1
[b]Open Europe, "Lost In Space: How the Cost of the EU's Galileo Project Has Skyrocketed", Briefing Note, October 2010, Available at http://www.openeurope.org.uk/Content/Documents/PDFs/galileo2010.pdf

and fixed price. A cost-plus contract (also known as a cost reimbursement contract) is based on the principle of paying the contractor for all of the incurred costs plus a predetermined amount of profit. Thus, if the cost of a project increases due to unforeseen circumstances, the buyer assumes the risk of a budget overrun.

The fixed-price contract, as the name implies, caps the total amount of the contract from the beginning. If the contractor successfully finishes all the project deliverables and manages to keep the costs low, his or her profit margins will increase. This time, the contractor assumes the risk of a budget overrun, as the client has already locked in a price limit.

Table 8.2 Costs for terrestrial mega projects

Project	Period	Cost in current year ($ billion)	Inflation adjusted cost in 2010 dollars ($ billion)	Notes
Large Hadron Collider[a]	1998-present	$10	$10.16	The reported operational costs per year are in the range of $1 billion
Chunnel tunnel[b]	1987–1994	$7.90	$15.17	Excluding operational costs and using an exchange rate of 1.7 from GBP to USD
Athens 2004 olympics[c]	2003–2004	$11.72	$13.52	Using an exchange rate of 1.31 from EUR to USD

[a]The Economist, "The Large Hadron Collider: The world's largest and most expensive experiment is up and running. Again." Nov 26, 2009
[b]Flyvbjerg, B., Bruzelius, N., Rothengatter, W., Megaprojects and Risk: An Anatomy of Ambition, Cambridge University Press, 2003
[c]Embassy of Greece, "Cost of Athens 2004 Olympics", Washington D. C., 2004, available at http://www.greekembassy.org/Embassy/content/en/Article.aspx?office=3&folder=200&article=14269

Although cost-plus contracts have a significant disadvantage of price uncertainty, they are an efficient instrument for managing highly complex projects that may have strategic importance. For example, during times of war or intense international cooperation, a government may assume the budgetary risks and make the overall project risks acceptable for the private sector contractors. However, as the tasks covered by contracts become more routine, the efficiency of the fixed-price contracts render them a better choice for minimizing the cost of budget overruns for the government.

Cost Management in the Private Sector

There is a widely held belief that government space programs are inefficient when it comes to managing the cost of space projects. Although there is merit in this line of argument, it is not a given that the private sector will achieve significant cost reductions by taking on the complexity of space projects with private capital, lean management practices and a less risk averse attitude. Observing the financial performance of the private space exploration initiatives discussed in Chap. 5 will be quite interesting in this regard.

Public-Private Partnerships (PPP) as a Cost-Sharing Mechanism

Public-private partnerships may look like an ideal way to control project costs and risks by pooling together the resources of the public and private sectors. The core idea behind the PPP model is to involve the public sector in the early stages of a project as an investor and the private sector as the designer and manufacturer. Once the project is up and running, the private companies start operating it, and, in theory, they reimburse the initial public investments (e.g., through royalties, tax payments, or free access to satellite data for government entities). When it works, the PPP model can increase the chances of success of a project by a transparent risk and benefit sharing mechanism. However, there are few successful examples of this in the space industry. Previous attempts, such as NASA's X-33 spacecraft, the European Galileo satellite navigation program and the Canadian Radarsat program all resulted in increased delays and yielded a smaller than expected return on investment for the government. On the other hand, NASA's Commercial Orbital Transportation System (COTS) program, which enabled SpaceX and Orbital Sciences to develop cargo delivery systems to the ISS, currently seems on track to prove a successful demonstration of how PPP can indeed work.

The success of a PPP implementation depends on the alignment of the interests of both the government and the private sector. Unfortunately, the fundamental differences between the priorities of the government and the private sector can easily get in the way. With relatively few competing companies in the space sector, and the government as a customer with an inelastic demand curve, there is little incentive in cutting costs.

Case Study: Does Anybody Actually Know the Cost of the ISS?

The International Space Station is one of the greatest technical achievements of our recent history. Built on the collective expertise of Russia, the United States, Europe, Canada, Japan and other international partners, its design, construction and operations are not just a testament to engineering but also to international cooperation. Although there have been some criticisms regarding the amount of science coming out of the ISS, one of the main points of contention has been its cost.

As part of his administration's strategy to establish U. S. dominance in space, President Reagan kicked off the program in 1984, aiming to counter the ambitious Soviet space station program. The original cost estimate was $8 billion (or $16.8 billion in 2010 dollars), and it was to be completed by 1992.[9] Dubbed "Space Station Freedom," the main purpose of the program was a show of force, rather than conducting space science experiments. In 1989, the George Bush Administration changed the plan slightly and positioned the station as a stepping stone for a subsequent human exploration mission to Mars. Partners from Europe, Canada and Japan were also invited to contribute to the station in exchange for usage rights.

The fall of the Berlin Wall the same year fundamentally changed the political landscape and forced the United States to revisit the primary purpose of the station. By inviting Russia to join the program in 1993, the United States aimed to achieve two ambitious goals – preventing the Russian know-how in space technologies from falling into wrong hands (such as North Korea or Iran) and decreasing the overall cost of the program by tapping into cheaper technology and launch costs in Russia. The station was also rebranded as the "International Space Station" to emphasize its new nature.

However, the diplomatic burden of an international partnership and the sheer challenge of merging space heritage from multiple countries dramatically increased the overall complexity of the program. By 2002, the cost of the ISS to the United States increased to about $30 billion.[10] Schedule slippages were a key factor in losing control of the budget. It was estimated that every month of delay cost about $100 million in the form of salaries, contractor overhead and sustaining engineering. In 1993, the completion of the in-orbit assembly was planned to take place in 2002. After nearly a decade of delays and 37 space shuttle flights, the ISS assembly was completed in 2011. International partners agreed to keep the station operational until at least 2020.

Since the ISS operations are ongoing, it is not possible to estimate its life-cycle cost with great accuracy. Furthermore, budgetary figures provided by NASA are generally not adjusted for inflation. As of 2010, NASA estimated that it invested more than $48 billion on the development and construction of the ISS since the beginning of the program.[11] Adjusted to 2010 dollars, this represents an investment of about $70 billion. However, this is only one of the cost elements.

Additional costs include (all in 2010 dollars):

- $53.3 billion (the cost of 37 shuttle flights at $1.44 billion each)
- $24 billion (the estimated cost of contributions from Europe, Russia, Japan, Canada and other international partners)
- $27 billion (estimated cost of operations about $3 billion per year until 2020)

Thus, the estimated life-cycle cost of the ISS is around $175 billion (in 2010 dollars), although the "official costing" provided by NASA and ESA for the International Space Station is $140 billion.

Notes

1. My thanks go to Charles de Gagne for this insightful comment.
2. Flyvbjerg, B., Bruzelius, N., Rothengatter, W., Megaprojects and Risk: An Anatomy of Ambition, Cambridge University Press, 2003.
3. Roger Pielke. R. Jr and Byerly, R., " Shuttle Programme Lifetime Cost", *Nature*, Vol. 472, April 2011.
4. United States General Accounting Office, "NASA: Lack of Disciplined Cost-Estimating Processes Hinders Effective Program Management," Washington D.C., May 2004, available at http://www.gao.gov/new.items/d04642.pdf.
5. NASA, *Cost Estimating Handbook*, Washington D.C., 2008, available at http://www.nasa.gov/pdf/263676main_2008-NASA-Cost-Handbook-FINAL_v6.pdf.
6. A variation of this method, costing by comparison, involves evaluating the proposals of competing bidders and comparing the cost information. For example, a space agency may issue a Request for Proposals without estimating the cost in advance, and then compare the cost of the received offers. Although such an approach can provide valuable information, an independent

cost evaluation is always advisable as the bidders may misinterpret the requirements or try to increase the cost arbitrarily.

7. Congressional Budget Office, "A Budgetary Analysis of NASA's New Vision for Space Exploration", Washington, D.C., 2004. Available at http://www.cbo.gov/sites/default/files/cbofiles/ftpdocs/57xx/doc5772/09-02-nasa.pdf.
8. NASA, Parametric Cost Estimating Handbook, available at http://cost.jsc.nasa.gov/PCEHHTML/pceh.htm.
9. *The Economist,* "A Waste of Space: The International Space Station Is About to Receive Its First Tenants", October 2000.
10. U. S. General Accounting Office, "Space Station: Actions Under Way to Manage Cost, but Significant Challenges Remain", Washington D.C., 2002, available at http://www.gao.gov/assets/240/235123.pdf.
11. United States General Accounting Office, " NASA: Significant Challenges Remain for Access, Use, and Sustainment of the International Space Station", Washington D.C., 2012, available at http://www.gao.gov/assets/590/589668.pdf.

Chapter 9
Putting It All Together: Assessing the Feasibility of a Space Venture

Building on the material we covered in the preceding chapters, we are now ready to put our knowledge of core economics and business concepts to use and structure a basic feasibility analysis for a new space venture. We will focus on on-orbit satellite servicing (OOS) as our case study.

OOS is defined as: "a service offered for scientific, security, or commercial reasons that entail an in-space operation on a selected client spacecraft to fulfill one or more of the following goals: inspect, move, refuel, repair, recover from launch failure, or add more capability to the system (see Chap. 5, endnote 6)." These clients can be satellites, space stations or other types of mission architecture elements (e.g., Mars mission modules). The focus of this case study is a venture that will develop a robotic capability to service in-orbit assets.

Market Overview for OOS

The main market for OOS in the short-run are the GEO satellites, although almost all types of satellites (larger than the 100-kg microsatellite class) are potential sources of demand. Currently there are about 1,000 operational satellites in orbit, and about 40 % of these are located in GEO. These satellites are primarily used for telecommunications, and with an average launch mass of 3.5 metric tons, they are much bigger than the satellites in other orbits.[1] Each year, about 25 satellites in GEO are retired; more often than not, this is due to the depletion of the on-board propellant used for keeping the satellite in its proper orbit.[2]

Space insurance data yield interesting facts regarding the economic value of OOS. The first year of a satellite's lifetime is also its riskiest. It is reported that between 2000 and 2011, 38 % of losses within the first year were due to launch failures and 45 % were incurred within the 2 months following a successful launch.[3] In other words, if a satellite survives the first 2 months in orbit, its odds of survival for the rest of its lifetime are quite high. Moreover, during the same time period, the average amount of annual losses directly attributable to an in-orbit mishap was about $300 million.[4] Thus, both satellite operators and insurance companies may be interested in a service that can potentially reduce total economic losses of about $3 billion incurred within the last decade. The presence of an OSS system could help reduce these losses and create significant value for many stakeholders in space business.

In addition to losing an orbital asset due to an in-orbit failure, satellite operators can also lose business revenue since a failed satellite may mean failed commitments to customers. Therefore, some satellite operators could be willing to pay more than the insured portion of the satellite for a servicing operation, since they will be able to keep their business commitments, avoid losing operating revenues and also keep their reputation intact as reliable operators. Furthermore, satellites nearing their nominal lifespan can get a boost in the form of refueling, and provide the option of a service extension to the satellite operators.

OOS may also have a very interesting impact on the demand for launch services. The overall reliability of satellites has increased significantly over the last decades, and it is not uncommon to see satellites with more than 10–15 years of lifetime. Surely, OOS would increase the reliability of satellite operations even further, suppressing the demand for new satellites and consequently new launches. However, OOS would also create additional demand for launch services, since many orbital replacement units and a significant amount of propellant have to be stored in orbit for servicing operations. The exact balance between this new demand and the loss of demand due to increased lifetime is not clear, and requires more research.

Defining the Business Model

In its simplest terms, a business model is a map that connects the key elements of our venture: the product and/or service that is being commercialized, the target market, the optimum method of distribution, the financing arrangements, etc.

The two fundamental forces of economic activity, demand and supply, will determine the likelihood of success of the new venture. More specifically, there is a need to make sure that there is a sufficient level of demand for the new products and services that can be addressed at a profitable price point. In the case of the OOS, service providers can decide to offer a "pay-as-you-go" product, where they can assess the complexity of a servicing mission and provide a quote to potential clients. An alternative business model is to offer a subscription-based service, whereby the service provider collects premiums from a broader pool of clients and then services the satellites without any extra charge. Other types of business models can also be envisioned.

Understanding the Supply Side

In broad terms, the supply side refers to all the activities required to bring a new product and/or service into the market. These activities include manufacturing, quality control, compliance with existing regulations, launch and all the other steps required to have an operational system in orbit. Although this is obviously a challenging undertaking by itself, it is not sufficient. There is also a need to make sure that the offering can compete with the existing products and services in the market through differentiation. There may be a need to offer higher quality, better performance, quicker service, lower price or a combination of the above in order to establish a competitive position.

In the case of OOS, one of the key challenges in this regard is to minimize the life-cycle cost of the required infrastructure without compromising the reliability of the servicing system. As a point of departure, the cost information from the Orbital Express mission of DARPA can be used for costing by analogy. This mission cost about $300 million in 2005 dollars.[5] Thus, a mission of similar complexity would cost about $350 million in 2012 dollars. Of course, a detailed cost analysis using parametric costing would provide more accurate estimates, including an upper and lower bound for cost. Not only the manufacturing costs but also the operations, maintenance and disposal costs should be included in the cost estimates.

Understanding the Demand Side

The demand side is all about the customer. Understanding the needs and priorities of the customer is essential for identifying the value of the products and services. Without this piece of information, it is simply not possible to determine the optimal pricing strategy. It is also important to realize that this perceived value may change from one customer to the other.

One possible strategy for this space venture is to start with servicing missions with relatively low complexity and clear value to the customer (such as extending the client satellite's lifetime by docking a refueling module). Once this particular segment is addressed, subsequent servicing missions can tackle more complex tasks, such as replacing on-board systems. Based on the preliminary market analysis, satellite operators who own assets in GEO are key customers. Governmental organizations, including the military, could also be a source of demand.

Is There a Match Between Demand and Supply?

The long-term success of the venture is critically dependent on the match between demand and supply. Given the long lead times for space missions and their inherent complexity, it is highly unlikely that the business will be immediately successful and start generating profits right away. Even though the initial cash flow may turn positive, this is not the same thing as profitability where the entire invested capital is recovered and profits start accumulating.

Risk Analysis

Using the risk matrix introduced earlier, we can analyze some of the primary risks of a commercial OOS venture. The potential impact of these risks and possible risk management strategies are discussed below.

Risk 1: Technical Complexity
Launch, rendezvous and servicing operations are highly complex. Capturing and manipulating a client satellite will inevitably raise the risk profile. As a mitigation strategy, in addition to extensive testing of the spacecraft, mission scenarios have to be simulated in great detail.

Risk 2: Cost Overruns
The target customers, commercial satellite operators, will be sensitive to the price. They always have the option to procure a new satellite or reshuffle their fleet to replace lost capacity in-orbit. Therefore a cost overrun may have very adverse consequences, as it may force the service provider to set the price higher than the value offered to the customers. Therefore it is critical to establish sound cost management practices and estimate the life-cycle cost. Cost sharing mechanisms, such as a public-private partnership model, may also be suitable.

Risk 3: Changing Market Conditions
By the time OOS is available as a commercial service, the needs and priorities of the customers may change. Companies such as SpaceX and Stratolaunch Systems may aggressively price their launch services and give satellite operators more incentive to launch new satellites instead of servicing them. Possible mitigation strategies include continuous monitoring of market conditions and changing course, if necessary. For example, if servicing GEO satellites is not feasible, focusing on a space debris solution may create new commercial possibilities.

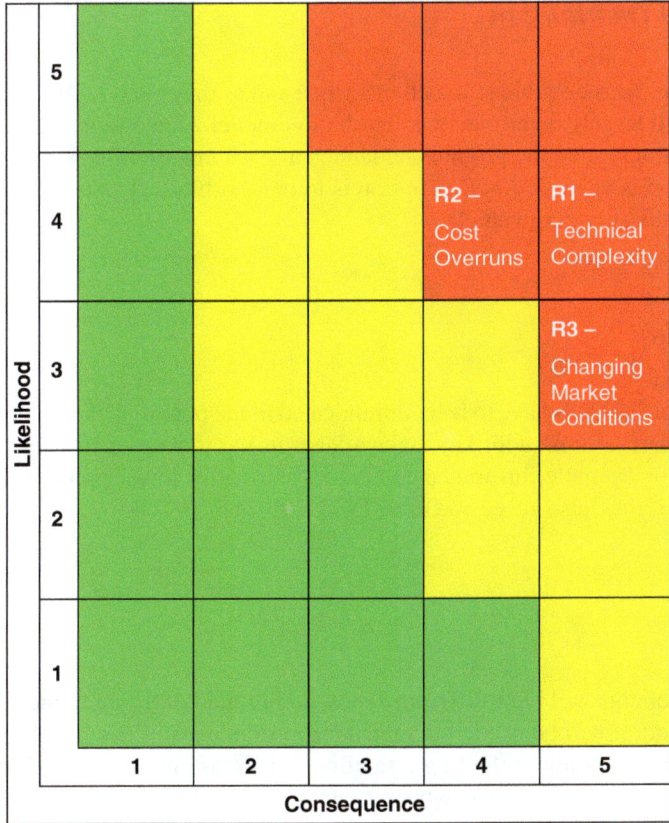

Naturally, the risks outlined above are interrelated. Any major technical problem will result in delays in project schedule and push the costs higher. This, in turn, will increase the price point at which the venture can be profitable.

Running the 5Ps of Marketing

The first step of a comprehensive marketing plan is to analyze the OOS venture with the "5 Ps of Marketing" framework.

Pricing

The "litmus test" of the feasibility analysis is to see if the service can be offered at a price point which offers a clear value to the client and enables the business to recover its costs and make a profit. In this regard, understanding the business model of satellite operators and their sources of revenue is essential.

Physical Distribution

OOS has to be accessible to the clients when and as they need it. The mission architecture will largely determine the distribution model. One option is to keep most of the infrastructure on the ground and launch a servicing satellite as soon as a servicing mission is needed. Another option is to build orbital "depots" to store the supplies and the servicing vehicle.

Promotion

The benefits of OOS have to be communicated to the potential clients. In this regard, building partnerships with key stakeholders in the industry can help promote the service. For example, insurance companies may offer lower premiums to satellite operators if they also subscribe to an OOS service.

Product

Different clients will have different needs, and instead of trying to meet all of them, a prioritization will be needed to match the system capabilities to the highest value services. For example, refueling a satellite to increase its operational lifetime may be seen as a routine activity, whereas capturing a malfunctioning satellite may be billed as an emergency operation.

Philosophy

By developing OOS capabilities, it is also possible to contribute to the environmental sustainability of space operations and better manage the threat of space debris. Active removal of space debris through OOS can boost the public profile of the venture and bring government contracts at the same time.

Notes

1. Union of Concerned Scientists, UCS Satellite Database, data as of April 1, 2012. The database is available at http://www.ucsusa.org/nuclear_weapons_and_global_security/space_weapons/technical_issues/ucs-satellite-database.html.
2. Foust, J., " The Space Industry Grapples with Satellite Servicing", *The Space Review*, June 2012, available at http://www.thespacereview.com/article/2108/1.

3. De Selding, P., "Falling Satellite Insurance Premiums Put Market at Risk of Major Upheaval", *Space News*, March 2012, available at http://www.spacenews.com/satellite_telecom/120302-falling-sat-insurance-premiums-market-risk.html.
4. Kunstadter, C., "View From the Leading Edge", presentation made at the World Space Risk Forum, February 2012, Dubai, available at http://worldspaceriskforum.com/2012/wp-content/uploads/2012/03/2CHRIS1.pdf.
5. Berger, B., "U. S. Air Force to End Orbital Express Mission," Space.com, available at http://www.space.com/4018-air-force-orbital-express-mission.html.

Chapter 10
Conclusions

Space is not a fringe industry of terrestrial economic activity; it is also not a land of fantasy where the fundamental laws of economics don't apply. Space is simply a medium, a vast universe full of resources to be explored, a vast ocean connecting stars and planets.

In this respect, history is about to repeat itself. Just like the early Greek and Phoenician settlements across the Mediterranean, or the first European settlements in the Americas, we will explore, settle and build new societies. The eventual success of these societies will depend on many factors, but economics will always play a key role.

Some segments of the space business are becoming mature, well-established industries. However, for sustained growth, there is a need to solve many challenges as outlined in the previous chapters. The competitive advantage of satellites as the ultimate higher ground should not be taken for granted. There is always competition from terrestrial substitutes. As the Iridium case highlights, the long lead times for space-based solutions may be just too long. A time horizon of 5–7 years may very well give competing business models ample time and opportunity to raise financing, develop products and fill the market niches with alternative technology that is better, faster, cheaper or more convenient. Whether it is air balloons or Unmanned Autonomous Vehicles (UAVs), there will always be other platforms that can provide comparable value – and perhaps at a fraction of the cost.

Companies such as SpaceX and Scaled Composites are rewriting the chapter on space business. Access to private capital, lean management practices and a much more aggressive management of risk accelerate the rate of change and spur many innovations. Innovation is not just about developing brand new technologies it's also very much about combining existing technologies in brand new ways.

The recently announced Stratolaunch Systems venture, the successful mission of SpaceX's Dragon spacecraft to the ISS and the partnership between Planetary Resources and Virgin Galactic to launch a space telescope as a first step for asteroid mining are just some of the recent developments that will shake up the industry. It's still very early to see if these ventures will succeed in securing the required funding and overcome all major technical and regulatory hurdles to achieve long-term success. What's clear, however, is that a new way of doing business is finally here: taking calculated risks, using private funding and combining the critical expertise of the private sector in a self-organizing way.

Until major cost reductions can be achieved, the most valuable space cargo will continue to be packages of information and not physical goods or passengers. Thus, a space venture doesn't necessarily have to be based on ownership of space assets. Before building a single piece of space hardware, entrepreneurs must explore possibilities to lease capacity or use existing satellite data products.

Today, space is serious business, but it's nowhere close to its true potential. Our perception of our surroundings is undergoing a major transformation: astronomers are discovering new exoplanets almost every week. The cosmos is teeming with new destinations to explore. In this sense, space is not just an Earth-bound market or industry, it's a host of technologies, know-how and new destinations; it's our vessel to explore the universe.

Top Ten Things to Know About Space Business and Economics

1. Cost is not equal to price. The space industry is moving towards fixed-price contracts and established companies need to adapt to this new market reality. Increasingly, products and services will be priced based on the value they offer to customers, regardless of how much they cost to build.
2. Space professionals need to better "market" space activities. They need to develop a more effective marketing mix and communicate the benefits of their projects clearly.
3. Government is still the biggest customer. As the industry matures, private sector is steadily increasing its clout. Companies such as SpaceX have proven that they can provide access to LEO. But it remains to be seen if this will result in any actual cost savings. Other projects such as Orbital Science's Antares/Cynus and Stratolaunch Systems must prove their viability to convert isolated successes into a true historic trend.
4. When it comes to risk and space, perhaps the greatest risk we are facing is not to invest in space-based capabilities at all. Without understanding planetary evolution by studying our planetary neighbors we may never fully grasp the reasons and consequences of climate change. Without keeping a close eye on NEOs, we can never be sure that Earth is safe from an imminent collision. Therefore, risk of inaction will always be more than the risk of space exploration.
5. A private space venture still needs to work with the government. As long as the government assumes the role of an anchor tenant or loyal customer, it is possible to generate revenues in the early years. This is clearly illustrated by the multi-billion dollar contracts that NASA has with SpaceX and Orbital Science to resupply the International Space Station.
6. We are moving from a global to a "planetary" economy. In the long-run, it is possible that there will be demand and supply centers in various points of our

Solar System. If, one day, there are human settlements on the Moon, Mars, and other planetary bodies, the destination problem will be completely solved and access to space will be a necessity. In the short term, the viability of point-to-point suborbital transportation on Earth will be a true test of a sustainable and profitable market for human spaceflight.
7. Portfolio diversification works quite well for the private sector, with its expertise in replicating many "missions" and managing financial risk through launch insurance and other means. However, the catastrophic risk of losing lives seriously limits the use of portfolio-based methods for human spaceflight.
8. Some of the challenges outlined in this book can actually be great business opportunities. For example, space ventures aimed at solving the space debris risks can be very interesting for governmental and private sector customers.
9. Conventional financing methods such as venture capital and private equity are generally not applicable to space ventures. Thus, identifying and implementing innovative ways of financing is critical for the success of a space venture.
10. It's not all about business. What we learn through space activities can be exactly what we need to benefit from the vast natural resources of the universe and further expand our civilization. Space can help us enormously in understanding Earth as a "system of systems" and developing its resources in a sustainable manner.

Printed in the USA
CPSIA information can be obtained
at www.ICGtesting.com
LVHW011114221123
764618LV00001B/33